"*Naturebot* is a beautifully written and researched exploration of the rapidly growing field of biomimicry and its implications for how we think of and live in the world. It is an important book, one which will begin a much-needed conversation about the social and ecological costs and benefits of this brave new nature."

— *Helena Feder, Associate Professor, East Carolina University*

"When in unfamiliar terrain, having an adept guide can make all the difference between getting lost and finding a way—between being frightened of new phenomena, and learning from surprise encounters. In this scintillating book, James Barilla serves as the steady, insightful guide readers need to navigate the wildly fabricated landscape of biomimicry in its myriad forms. *Naturebot* is the story of a place that involves us, a living topos which is rapidly becoming everywhere."

— *Christopher Schaberg, Dorothy Harrell Brown Distinguished Professor of English, Loyola University New Orleans*

"A timely, fascinating take on the intersection of biology and technology, *Naturebot* will change your view of the future. With humor and insight, Barilla takes us on a compelling journey into a new kind of wildness—one that will appeal to scholar, student, and casual reader alike—a technological terrain that raises provocative questions about the connections between nature, science, and art. What Barilla finds in his encounters with unconventional nature is at turns absurd, frightening, and hopeful—often all at once. His message is one we cannot hear too often: our future will be what we make it, and in fact we are making it right now. It would be good for us to pay attention and choose wisely. *Naturebot* goes far in showing us how."

— *Paul Bogard, author of* The End of Night *and* The Ground Beneath Us

NATUREBOT

Naturebot: Unconventional Visions of Nature presents a humanities-oriented addition to the literature on biomimetics and bioinspiration, an interdisciplinary field which investigates what it means to mimic nature with technology.

This technology mirrors the biodiversity of nature, and it is precisely this creation of technological metaphors for the intricate workings of the natural world that is the real subject of *Naturebot*. Over the course of the book, Barilla applies the narrative conventions of the nature writing genre to this unconventional vision of nature, contrasting the traditional tropes and questions of natural history with an expanding menagerie of creatures that defy conventional categories of natural and artificial. In keeping with its nature writing approach, the book takes us to where we can encounter these creatures, examining the technological models and the biotic specimens that inspired them. In doing so, it contemplates the future of the human relationship to the environment, and the future of nature writing in the 21st century.

This book will be of great interest to students and scholars of biomimetics, environmental literary studies/ecocriticism, and the environmental humanities.

James Barilla is associate professor in the English Department at the University of South Carolina, where he teaches creative nonfiction and environmental writing in the MFA program. He is the author of two other books about the human relationship with the natural world: *My Backyard Jungle* and *West with the Rise.*

Routledge Environmental Humanities

Series editors: Scott Slovic (University of Idaho, USA), Joni Adamson (Arizona State University, USA) and Yuki Masami (Aoyama Gakuin University, Japan)

The *Routledge Environmental Humanities* series is an original and inspiring venture recognising that today's world agricultural and water crises, ocean pollution and resource depletion, global warming from greenhouse gases, urban sprawl, overpopulation, food insecurity and environmental justice are all *crises of culture*.

The reality of understanding and finding adaptive solutions to our present and future environmental challenges has shifted the epicenter of environmental studies away from an exclusively scientific and technological framework to one that depends on the human-focused disciplines and ideas of the humanities and allied social sciences.

We thus welcome book proposals from all humanities and social sciences disciplines for an inclusive and interdisciplinary series. We favour manuscripts aimed at an international readership and written in a lively and accessible style. The readership comprises scholars and students from the humanities and social sciences and thoughtful readers concerned about the human dimensions of environmental change.

Weather, Religion and Climate Change
Sigurd Bergmann

Monsters, Catastrophes and the Anthropocene
A Postcolonial Critique
Gaia Giuliani

Food for Degrowth
Perspectives and Practices
Edited by Anitra Nelson and Ferne Edwards

Urban Ecology and Intervention in the 21st Century Americas
Verticality, Catastrophe, and the Mediated City
Allison M. Schifani

Naturebot
Unconventional Visions of Nature
James Barilla

Trees in Nineteenth-Century English Fiction
The Silvicultural Novel
Anna Burton

Imagining Climate Engineering
Dreaming the designer climate
Jeroen Oomen

For more information about this series, please visit: https://www.routledge.com

NATUREBOT

Unconventional Visions of Nature

James Barilla

Routledge
Taylor & Francis Group
LONDON AND NEW YORK

First published 2021
by Routledge
2 Park Square, Milton Park, Abingdon, Oxon OX14 4RN

and by Routledge
52 Vanderbilt Avenue, New York, NY 10017

Routledge is an imprint of the Taylor & Francis Group, an informa business

British Library Cataloguing-in-Publication Data
A catalogue record for this book is available from the British Library

Library of Congress Cataloging-in-Publication Data
A catalog record has been requested for this book

ISBN: 978-0-367-56751-4 (hbk)
ISBN: 978-0-367-60779-1 (pbk)
ISBN: 978-1-003-10047-8 (ebk)

Typeset in Bembo
by codeMantra

To my parents, Jerome Barilla and Dorothy Laible, for inspiring me to explore the world

CONTENTS

ACKNOWLEDGMENTS

So many people helped guide me on the journey that resulted in this book, and one of the chief pleasures of writing it was meeting so many fascinating and generous people along the way. With their time, their expertise, and their stories, they pointed me in the right direction, sustained me with food-for-thought, and opened new vistas for me to explore. I am so glad our paths crossed. I owe a special thanks to Ingrid Mari-Anne S. Gaup, who graciously allowed my family and I to visit her home. Erika Larsen, Inger Marie Nilut, Bruce Forbes, David G. Andersen, and Tracie Curry all offered invaluable advice on reindeer herding and the logistics of travel in northern Scandinavia. Jason Matheny, Isha Datar, Nick Halla, Cor van der Weele, and Claus Driessen helped shape my understanding of the history and context of cultured meat and cellular agriculture. For their generosity with their time and their indulgence of my questions, my genuine thanks to Maureen Leary and the many volunteers, trainers, and staff at PAALS. I am deeply grateful for my memorable visit to The Burrowes at Salvington Lodge and to everyone who made it possible, and to Jennie Deese and Eve Anthony and everyone at the Athens Community Council on Aging (ACCA) Bentley Center for Adult Day Health for graciously welcoming my visit. James Bellingham, Ioannis Ieropoulos, Laura Fogg Rogers, and Elliott Hawkes all helped introduce me to the field of biomimetic and bioinspired robotics. A special thanks to Marco Valtorta, who helped me with the technical aspects of some of the chapters; to David Robertson for the inspiration to engage with nature and culture creatively; and to Susan Courtney and Bob Bohl.

Thank you to my colleagues in the English Department at the University of South Carolina, especially Nina Levine and Greg Forter, who provided timely advice and support. The Provost's Office at the University of South Carolina provided essential funding for travel to northern Norway and Finland. The Dean's Office of the College of Arts and Sciences at the University of South

Carolina also provided crucial support to allow me to conduct research in the United Kingdom. I am deeply indebted to Wendy Strothman and Lauren MacLeod, who patiently read through multiple early versions of this work and brought clarity and purpose to its ultimate form. I wish to thank Oindrila Bose, Annabelle Harris, Rebecca Brennan, Grace Harrison and the production team at Routledge for shepherding this project through the editing process so deftly while diligently addressing all my questions of style and substance. Scott Slovic, Joni Adamson, and Yuki Masami offered vital guidance, support, and feedback. Their appreciation of the unconventional nature of this project was essential, and I am truly thankful for the opportunity to work with them.

And of course, my deepest gratitude to Nicola, Brook, and Beatrice for accompanying me on the journey. It wouldn't have happened without you.

INTRODUCTION

The biomimetic trail

One autumn afternoon, I headed out on a nature walk through the coastal forest with the South Carolina naturalist and television raconteur, Rudy Mancke, along with a crowd of eager listeners. To start, we gathered in the shade of a live oak and looked out through the canopy to the hazy cirrus lounging over the waters of Winyah Bay as he described the ecology of where we stood. There would be a degree of serendipity on this ramble through the oak and magnolia and palmetto stands of this coastal forest—nobody knew how far we'd get, or really which direction we'd take. Discovery would lead the way.

The first thing he happened upon was a strip of hide. He picked it up and stretched it out to display its length, then plucked with a fingernail at its sharp and peculiar scales. *Not a snake,* he said. *Longnose gar.* From there, we learned about the distinctive natural history of this ancient fish, the needle nose design of its jaws for snapping fish sideways, the way it lurks near the surface in stagnant water, using a special air bladder to breathe air. He speculated about the ecological connections that might have led to it winding up on shore, including what appeared to be a couple of bullet holes. Then we ambled a bit further before investigating burrows dug by cicada killers—giant specialist wasps that prowl around the canopies of trees and shrubs, hunting their prey, which they paralyze and drag into a subterranean chamber to feed to their offspring. We followed lines of ants ascending dead branches that were wattled with turkey tail fungi. We inspected the colonies of tiny resurrection ferns clinging to the shady horizontal boughs of the live oaks; they're known for their ability to wither brown as leather during a drought and spring back green after rain. Mancke is known for his keen eye, for racing to a spot where there appears to be nothing and emerging with a snake—I was told he'd discovered a juvenile rattlesnake earlier by the entrance. He's also known for the breadth of his knowledge of natural history, not one thing alone but the intricacies of the way things fit together in the natural world, human and otherwise.

We didn't travel far on that loop around the grounds, but it was an expansive journey nevertheless, both a practice of discovering what's out there and a performance for the crowd. What we had at the end of this walk was not an appreciation of a single feature of the local ecology, but a sense of the complexity of the ecosystem as a whole. We'd discovered patterns and connections running through the landscape from cloud to rainfall to fern. We had what amounted to a narrative, the story of what we'd seen and touched and identified, and we also had a kind of list, based on serendipity and keen observation, that wasn't itemized according to any kind of hierarchical taxonomy—the fungi and the gar were both of equal interest.

This walk left me pondering more than just the Lowcountry natural history—it felt like an extended metaphor for a journey of my own. I wanted to create a biomimetic version of this nature walk, an extended metaphor for the exploration of an expanding menagerie of creatures that aren't easily categorized, that blur the line between nature and technology. These robotic creatures trot on four legs like horses, slither like snakes, swim like pike, and swarm like bees; what follows is their unconventional natural history, written as if it were a journey whose purpose is to capture the breadth of their possibilities and the complexities of the connections between them.

Before we set out, we need to read up on where we're going; I always like to leaf through the pages of field and trail guides before a walk, getting a little deeper into the details of flora and fauna, and the trail highlights and conditions, even when they get a bit technical. We need to define the terms of this engagement with the landscape in other words, as if we're packing our intellectual gear. Water bottle, trail mix, first aid kit, sun hat, binos...on this journey, our pack will be full of terms and ideas instead, like *mimesis* and *agential realism*. But don't get too bogged down with the packing; this book is going to be about the experience of walking, not what we're carrying with us.

Naturebot is about the technological mirror we hold up to the natural world, a paradigm known as biomimetic (or the more flexible "bioinspired," as some in the field prefer) technology. The term biomimetics was first coined in the 1950s by the biomedical engineering pioneer Otto Schmitt, whose early research focused on emulating neural pathways in squid. A math and physics whiz who also trained in zoology, Schmitt's method involved investigating an anatomical system thoroughly enough to develop a model of its structure and function, and then reproducing an artificial version that might ultimately be useful—he held a number of patents in electronics. As a professor at the University of Minnesota, he worked in the interdisciplinary field of biophysics, applying theories and methods from physics to biological problems. But he saw in biophysics the possibility of another approach that he felt was being neglected: biologists could utilize their ideas and techniques to understand problems in physics and engineering. Biomimetics, then, was born of addition, in the broadening of possibilities through this back-and-forth, this consideration of the flip side, this representation of one as the other. A remembrance of Schmitt's life and career describes him as having "a talent for connections," for identifying common threads between seemingly unrelated phenomena and disparate academic disciplines.

From the start, then, this transdisciplinary practice was not going to lend itself easily to a single definition, and indeed the terminology for what's happening here has proliferated over time. In a review of the field published in 2006, Julian Vincent offered a catalogue of terms to be treated synonymously: "'biomimesis', 'biomimicry', 'bionics', 'biognosis', 'biologically inspired design,'" all "implying copying or adaptation or derivation from biology." We might add "artificial ethology" to this list, as the study of animal behavior by technological means has come to be known. Writing in the inaugural volume of the journal *Bioinspiration & Biomimetics*, which also appeared in 2006, Yoseph Bar-Cohen defined biomimetics as representing "the study and imitation of nature's methods, designs, and processes." In a later book, he described biomimetics as a "field of science and engineering that seeks to understand and to use nature as a model for copying, adapting, and inspiring concepts and designs."

These suggest a mechanistic and utilitarian paradigm, but Janine Benyus had used the term "biomimicry" in 1997 to advocate looking to the natural world for design ideas in order to "live in harmony with Nature." Her approach to "nature-inspired design" was to consider principles drawn from ecology that might guide all sorts of applications, including the design of buildings or urban landscapes, meant to assuage our hunger for "instructions about how to live sanely and sustainably on the Earth." We might think of it as biomimetics with an ethical dimension, a vision of how humans *should* create technology, with Nature as our mentor.

Vincent, however, suggested that this burgeoning field was still more ad hoc than codified; it still lacked a core model for its own operation, a "framework" for the process:

> Although it is well known that design and engineering are rendered much easier with the use of theory, in biomimetics, every time we need to design a new technical system we have to start afresh, trying and testing various biological systems as potential prototypes and striving to make some adapted engineered version of the biomimetic device which we are trying to create...Some form or procedure of *interpretation* or *translation* from biology to technology is required.

Interpretation and translation sound like potentially humanistic endeavors, especially since nature interpretation is a more formal description of what naturalists are commonly trained to do. But where do the humanities come in here? Why aren't humanistic disciplines trespassing more frequently across this field? Partly, it has to do with the checkered history of *mimesis*, a term for the act of mirroring or imitation that takes us back to ancient Greece. Plato famously denigrated *mimesis* as producing flawed imitations of our already flawed and hazy perception of the real world, but Aristotle recuperated the term. Art that mirrored life could indeed capture some of the existential truth of being in the world. The vexing question wasn't whether humans copied or imitated nature, it was the assumptions that surrounded that practice, and what they revealed about the value of the original, the accuracy of the imitation, and the act of mirroring itself. What can we make, for example, of a mime's performance? What about the proliferation of copies in the mechanical age? Perhaps more fundamentally, how can we claim to know anything about the actual world, outside of the language we use to represent it?

As the world of living beings and material objects grew inaccessible in humanist circles, the need for mimesis as a practice for engaging with it declined. There were those, like the anthropologist Michael Taussig, who still saw mimesis as relevant, and whose description of it in his 1993 book, *Mimesis and Alterity*, feels prescient today:

> I call it the mimetic faculty, the nature that culture uses to create second nature, the faculty to copy, imitate, make models, explore difference, yield into and become Other. The wonder of mimesis lies in the copy drawing on the character and power of the original, to the point whereby the representation may even assume that character and that power.

The nature that culture uses to create second nature doesn't sound that far off from the definitions of biomimetics offered by engineers, albeit with an additional sense of the qualitative implications of being drawn into otherness. And yet, as the classicist Stephen Halliwell observed in 1990, "In the eyes of many, mimesis has the status of a venerably long-lived but now outmoded aesthetic doctrine—a broken column, perhaps, of an obsolete classical tradition." In 2007, the literary critic Timothy Morton identified "ecomimesis" as a pervasive fallacy to be found in most conventionally "green" writing about Nature. Such writing suggests immersion in the natural world, but in actuality reinforces assumptions about human estrangement and separation from ecological systems in all their impurity. There is no objective Nature out there, distinct from us, and hence no way to create a second nature through mimesis. Nature is already second nature.

Call me a contrarian, but when I see resolute positions taken in opposing directions, I tend to think something worthwhile is happening between them. What I want to embrace on this journey is the tension and synergy that comes through addition, through the conjoining of *bios* and *mimesis*. The wheels of theoretical discourse have continued to turn, such that we've reached a kind of convergence point between renewed humanistic interest in engaging the material world of tangible bodies and scientific interest in mimicking embodied systems. Interest in the "mimetic faculty," for example, has been revivified with the identification of mirror neurons which are said to define our own behavior via the human tendency to imitate others. They've even been described as part of an "as-if body loop system." Seeing an expression or a behavior in someone else triggers these neural pathways; we respond in kind, or with kindness—some researchers theorize that empathy arises via these mechanisms of neurobiological mirroring. Mirror cells cause metaphysical questions of the engagement between self and other to appear in the pages of scientific journals; they connect mimesis once more to alterity, via neurobiology. We are, it seems, mimetic creatures once again.

More broadly, we might view these developments as manifestations of what has been termed "new materialism," with its emphasis on the value of phenomena that are, as the roboticist Rodney Brooks once put it, "grounded in the physical world." But I like to think of what we're up to here as partaking of the theologian William Schweiker's description of mimesis in 1988: "Mimesis is best understood not as iconic copying but as the praxis of figuration. Mimesis is a way to speak of those forms of praxis in and through which we participate in a meaningful world." Biomimetics is a practice, then, which allows us to participate in the natural world by creating meaning through figuration. It bears some kinship to what the digital media professor Ian Bogost has described as "metaphorism," and to the practice of philosophical inquiry that the physicist-turned-philosopher Karen Barad has called "ethico-onto-epis-temology," in which questions of ethics and existence and knowledge are inextricably linked. That feels like it might also describe the multidimensional practice of nature writing, the way meaning arises from watching a monarch settle on a sprig of goldenrod, the way Aldo Leopold manages to touch on each of these dimensions while writing about listening to a chickadee at dawn, paying attention to what's out there, and what's happening in his own interior landscape, and the porosity of the boundaries between them. There's mirroring going on in Leopold's prose, a curiosity about natural history and a fidelity to biotic detail. But there's inspiration too, new insights about the human place in the natural world that led to the development of Leopold's "land ethic," a paradigm that challenged the orthodoxies of the time. A biomimetic take on nature writing performs similar work, crafting metaphors that connect the experience of learning about the natural world to the experience of being immersed in it, and in turn raising the possibility of a more informed ethics.

At the heart of this biomimetic practice, this understanding of the one via the other is a figurative gesture. It's conventionally called a model or a representation, but we might think of it as metaphor as well, and therefore consider the humanistic contribution to biomimetics as the creation and study of the metaphors we make of the natural world. A humanities-inflected biomimetics acknowledges its own limitations, following in the footsteps of the philosopher Graham Harman, who in *Guerilla Metaphysics* makes the case that only certain kinds of metaphor succeed. "Some metaphors *do* fail—and the way they fail is by never pushing us toward the world of unified things, remaining frozen instead on the layer of inert qualities." Harman demands wholistic gestures, more than mere "handshake agreements on shared qualities." Any technological representation of an ant, for example, that could be just as accurately portrayed by modeling its constituent properties won't suffice. The metaphor has to represent the dynamic, ultimately unpredictable character of a creature in its subjective experience of space and time. Barad offers an additional caution with a metaphoric take on diffraction: the way light scatters and leaks away without ever reaching the aperture in our eye, she suggests, is indicative of the way our experience of what's out there is always limited in detail. Diffraction prevents any attempt to mirror what's out there with absolute clarity and certainty.

And yet, this biomimetic practice still depends on scientific discoveries to recognize what's going on out there. Without the vast store of scientific knowledge that surrounds both the biotic and technological realms, knowledge generated by the practice of scientific inquiry, the possibilities for mimesis become simplistic rather than wholistic. Mimetic complexity arises from the interaction of different paradigms. Rather than dismissing conventional approaches to science and engineering, metaphorism would actually seem to complement them, assembling their mechanisms and patterns and properties into increasingly wholistic representations.

We can find an echo of these questions about the accuracy and goals of mimetic praxis in the current use of the terms biomimetic and bioinspired. The two are often deployed interchangeably or in tandem, as in the title of the journal, *Bioinspiration & Biomimetics*. But as we'll see, they also suggest two very different visions of mimesis, one of which desires creative license, while the other values faithfulness to the original. Bioinspiration sees evolution as the starting point for innovation; biomimetics views it as the finish line for knowledge of how the natural world works. It's really a debate over what Taussig calls "mimeticity," the degree to which one thing mirrors the other. I'd encountered something similar in another arena: ecological restoration, where one of the fundamental points of contention was the degree to which the design of restored ecosystems must maintain "fidelity" to what the landscape looked like before humans arrived to degrade it. In the case of bioinspiration and biomimetics, however, the two terms often coexist. The interdisciplinary and transdisciplinary nature of the practice itself seems to allow both fidelity and creative license.

The result is an evolving ecosystem populated by technological creatures, all displaying varying degrees of mimeticity. If it exists in Nature, then someone is probably trying to create a technological doppelgänger. A robotic seal is a metaphoric take on what it is to be a seal in the physical world, built to be experienced by us in that world—it's an embodied representation or metaphor that illuminates in multiple directions. It's a representation of the natural history of pinnipeds. And a representation of the human relationship to other species. And a reflection of the strategies that interpret and translate, strategies that suggest both the possibilities and limitations of creating this second nature. The same might be said of the bat ears created by the British artist Alex De Little, inspired by three-dimensional representations of horseshoe bat ears developed by the mechanical engineering professor Rolf Müller and meant to be worn by human listeners to inspire a sense of "sonic empathy" across species lines. There's a kinship here, a lineage in this assemblage of scientific, robotic, artistic, and theoretical approaches, one that nature writing also shares.

Think of this book as a walk through a technological landscape in the company of a writer who routinely scans the treetops for songbirds and peers into crevices for creepy crawlies, searching for clues about the relationship between human nature and the natural world. Let's saunter in Thoreauvian fashion through this expanding synthetic ecosystem, mirror neurons firing on all cylinders, and see what's evolving there. It's not a literal account, in the manner of traditional nature writing, not the daily journal of life along the Pacific Crest Trail, not even a diary of the sort the naturalist Charles Darwin kept of his voyage aboard the Beagle. It's a performance of those documentations. This book is itself a biomimetic artifact of sorts, an extended metaphor of a trail that doesn't literally exist in temporal or spatial terms, but instead is a literary model meant to transcend the words on the page, meant to lead us, imaginatively, out there, where the wild things roam. It's also an invitation to view technology in a new light, to see it, not as a potentially threatening terra incognita or a source of alienation, but as full of splendor and possibility, too. Why not learn to deepen our attention to what surrounds us, as we might if we were out hunting morels or matsutakes, finding only twigs and leaves and dappled sunlight, and then suddenly seeing their distinctive shapes come into focus? The kinship between technology and nature is also percolating just beyond what we normally perceive.

Instead of solely looking *to* Nature for lessons, I wanted to look *from* Nature. I wanted to consider both, from the liminal space between the two. It's uncomfortable, for a nature writer, to confront the indelible presence of technology in what would otherwise be a quintessentially "natural" encounter with another species—but that's what makes it fascinating. How do we reckon with our technology—that is a fundamental, if not *the* fundamental, question for nature writing in the 21st century, since the defining issue of our time is the planetary climate crisis of our own making. One familiar response is to turn away from technology, an impulse already apparent in "no impact" movements that include avoiding plastic or refusing to travel by plane. This resistance is longstanding, as the biomechanics researcher Steven Vogel encountered in attempting to find common ground for *Cats' Paws and Catapults: Mechanical Worlds of Nature and People.* "In getting together the material for this book," he writes, "I've repeatedly bumped into an anti-technological literature—nature-worshipping, engineer-bashing tracts." Of course, there's the opposite, too, the faith in "technofixes" that offer the possibility of transcending our ecological ills, perhaps even by creating virtual realms where we don't go outside at all.

But is there an alternative route, a paradigm for the engagement of human creative endeavor and ecology? Not a solution necessarily, but a practice for seeking one? What is the nature writing of that movement? What is the nature writing, for example, of the colliding visions of Nature that are accompanying the emergence of "cellular agriculture," of biomimetic "beef" grown in labs in hopes of cutting greenhouse gas emissions and saving the planet from conventional carnivores? What can biomimetics reveal not just about Nature as subject, but also about nature writing as literary practice, by applying the tropes and motifs and narrative hallmarks of the genre to these biomimetic creatures that aren't easily categorized, that blur the line between nature and technology and in doing so illuminate the struggle to articulate a nonbinary nature? Can biomimetic nature writing find a paradigm for the interweaving of human activity and forces that lie outside of our understanding and control? In the current climate, that search has become more urgent than ever.

We might consider this mimetic journey in terms of the broader context that is playing out in record-breaking heat, melting glaciers, whirlwinds of fire and rain worldwide: the climate emergency, and how it is reshaping our sense of human engagement with the world we share with other species. When Bill McKibben published *The End of Nature* in 1989, he described a fundamental shift in our relationship to the planet—we could no longer maintain the illusion of Nature as separate from ourselves and our activities. The romanticized Nature that was transcendent, that was bigger and wilder than us, had been supplanted by something else. The Anthropocene is one way of describing the terms of this transformation, although describing the times we live in doesn't quite capture the profundity of what's changed with our vision of Nature itself. This is a crisis of identity as much as it is an ecological crisis—how are we to reconcile our sense of what it means to be human with this sense of planetary responsibility? What is the vision of this new Nature? How do we make sense of its meaning, and put that understanding into practice?

This struggle to articulate where we are with Nature feels pervasive as the signs of climate change take hold—we can see the theorist Bruno Latour reconfiguring the Gaia hypothesis as an extended metaphor for "entangled" relationships, and Donna Haraway, famous for complicating the conventional divisions between technology and biology in a "cyborg manifesto," now describing our contemporary circumstances as a bundle of connections between different species, all of us "working through" the challenges we face through the relationships we forge. From the "object-oriented ontology" of Harmon to the "agential realism" of Barad, what we find in the shadow of the Anthropocene is a focus on the tangled relationships between the living and the nonliving in the real world. What lies between different "agents" is what defines them.

That's the broader context for this biomimetic journey, the fact that we are at an inflection point with our sense of who we are as a species and how we fit with the world around us. For me, there's something about reading conventional nature writing in this context that doesn't feel accurate to where we are—not the form itself necessarily, not the hallmarks of the practice, but the assumptions built into the subject matter, the vision of what constitutes the human relationship with Nature. I wanted to start this nature walk with what I was experiencing, which was the fact that technology was indelibly present—that was the assumption from which to proceed in the Anthropocene. From there, I began to look for a paradigm that considered technology and Nature as "entangled," to use that term from philosophical circles. That's how I began to consider biomimetics—not solely its mechanistic and utilitarian aspects, but as a broader paradigm for how humans relate to the rest of the natural world.

Biomimetic nature writing is not a prescription for how to heal the planet; it differs from biomimicry in that regard. A biomimetic narrative is an exploration of the ecotone, the transitional landscape between two seemingly irreconcilable visions of Nature. One idealizes the natural world by repudiating technology: getting closer to Nature, whether via a sojourn in the wilderness or through a lifestyle purge of modern conveniences, means leaving behind the contaminating influences of human invention. Nature is our refuge from technology—that's one familiar story.

At the same time, this idealization has long been shadowed by a canonical skepticism over the "red in tooth and claw" implications of natural selection. Take Joyce Carol Oates's denunciation after enduring a fainting spell while out for a walk through the green and twittering countryside: "Nature has no instructions for mankind except that our poor beleaguered humanist-democratic way of life, our fantasies of the individual's high worth, our sense that the weak, no less than the strong, have a right to survive, are absurd." Or take a more recent, technologically inflected example from *The New Yorker's* Michael Specter, interviewed on NPR's Fresh Air about the potential use of gene-editing techniques in mice to prevent the transmission of Lyme disease: "The idea that somehow things out in nature are great and that if we mess with them the situation will be worse is sophistry...the amount of death and cruelty that exists in the natural world is unspeakably huge." Don't look to Nature, in other words, for lessons about how humans should behave with technology. Technology is our refuge from Nature—that's the other familiar story. Technology is what protects us from the natural world; it's how we defy, or at least forestall, a reckoning with natural selection.

The chapters in this book are about technology that is overtly, almost ostentatiously, invoking the natural world in its design. They represent the experience of Nature that comes *through* technology, beginning with the first chapter, which considers the movement of plants and horses as models for robotic mobility. The walk continues in Chapter 2 with a comparison of the way bats experience the world and how that might be imitated with a bat-inspired sonar head. Chapter 3 takes us into proximity with swarms of bees, ants, and robots. The next stage of the walk, Chapter 4, features an encounter with trailside companions both canine and robotic. Finally, in Chapter 5, we traverse the pastoral terrain of raising animals for consumption, starting the old-fashioned way with reindeer herding in northern Scandinavia, before heading into the industrial landscape of cellular agriculture and biomimetic burgers.

What vision of our place in the natural world emerges from this practice of observation and metaphor-making? That to me is the crux of the matter in a time of anthropogenic planetary crisis: are we building something unnaturally new here, leaving the biotic world behind like the husk of some kind of chrysalis? Or are we still hitched to everything else in the same tapestry, the same universe, as John Muir once put it. Can we transcend Nature, or are we always in this process of engagement, the one mediating the other?

We've assembled our intellectual gear, weighing its worth, stowing it and strapping it; even if you try to pack light, it's always heavier than you think.

Now it's time to hoist the pack and amble down the trail that leads into robot country.

1

TRAILHEAD

Quadrupeds and Plantoids

There's an old adage when it comes to planting bamboo: *in the first year it sleeps, in the second year it creeps, and in the third year, it leaps!*

The bamboo I planted in our South Carolina yard, *Phyllostachys nigra*, or black bamboo, held pretty much to schedule. When it arrived on our doorstep, it had been snipped down to a miserable looking wand dangling from a clump of roots and wrapped in wet newspaper. It didn't look like much, in other words. It didn't look like the photos I'd seen of stalks towering 30 feet in the air. Although they call it running bamboo in horticultural circles, I wasn't convinced. I'd arrived in our subtropical location from temperate latitudes, and it all seemed a bit hyperbolic to me. Run? It's a plant. It can't run. It can't leap. It can't even move.

For two years it was well-mannered. Each spring, it sent up two or three green stalks, which gradually ripened to ebony over the course of the year. In the meantime, however, I'd witnessed something that made me a bit uneasy. There's a technical college campus not far from us, and in the back of the parking lot is a stand of this very bamboo. More of a thicket, actually. The area is rectangular in shape, and in this rectangle is black bamboo. Nothing but bamboo, except for a parking space someone hacked out and covered with gravel. From the gravel, I observed culms pushing up like the tips of spears, bent on reclaiming the space.

This rectangle of bamboo was about the size of our entire yard. Including the house.

I thought I'd taken this appetite for expansion into account by planting the bamboo in a sliver of dirt in front of our garage, surrounded by asphalt, concrete, and brick pavers on all sides. Surely, it was trapped inside this hardscaping?

Year three: what is that thing coming up out of the middle of the lawn? It looks like a zombie's forearm, complete with sloughing skin. There's another one. And another one over there. What the…it's the zombie apocalypse!

Underground, in a hidden realm known technically as the rhizosphere, the bamboo had been busily responding to signals in the environment. The ground beneath us is wrapped in rhizomes and roots, threaded with fungal hyphae. Some delve down, anchoring themselves with a tap root, as you'll know well if you've ever tried to eradicate a dandelion. Some go searching horizontally, erupting far from the source. Some are thick and tough as an elephant's foot. Others are microscopic, fine root hairs sipping from the soil.

Beneath the pavers, the runners had burrowed, pushing their way invisibly through the grit and sand until they reached the other side of the patio. Ten feet. Then another seven or eight feet to reach the greener pastures of the lawn. Then up into view. It appeared, from above ground, to have jumped the concrete boundaries—to be growing by leaps and bounds, just as the old adage predicted. Within a day of its emergence, it rained, and by the next day I wasn't looking down at these new culms—I was looking up.

If you'd asked me, I would've said the bamboo had moved. I didn't put it over there—I put it over here. But of course, it hadn't moved in a conventional sense. Not by locomotion. The runners covered the distance incrementally; instead of legging it like a dog chasing a squirrel, they expanded at the tip of these underground probes, sensing their way past the obstacles I'd set, charting a route. Cell by cell, the plant had literally grown to its destination.

I admired the ingenuity of the effort, although I wasn't keen on the result. One of the disadvantages of movement by cellular growth is that you can't run away, even if you're a running bamboo. Not even if your erstwhile benefactor is standing there with a pair of loppers.

~

Thinking about mobility, about how to get from one place to another—that's a good place for us to start, since we're setting out on a journey ourselves, metaphorically speaking. At the edge of this bamboo thicket lies the trailhead, the spot where there's a placard with a map and notices from authorities, some warnings, some invitations to look out for some feature. "You are here," it says on that faded, cobwebbed topo sheaf, a red pushpin, a dab of marker on the concentric lines. Or maybe our first sign harkens back to the one Frank Lutz, a curator at the American Museum of Natural History, posted in upstate New York in 1925, at the start of a nature trail that is said to be the very first of its kind. The sign read: "A friend somewhat versed in Natural History is taking a walk with you and calling your attention to interesting things." That, Lutz wrote, was the spirit of the nature trail.

Sign the register and put the annotated sheaf back in its wooden box. There's a blue blaze on the trunk ahead. You can see the path wending its way, outward bound. It's time to get moving.

Movement is the plot device, the underlying arc or storyline, for much of nature writing—going out into the wilderness, or even more fundamentally, just going outdoors, implies a starting point and a destination. The scope of the journey varies—a touchstone of mine is Aldo Leopold's early morning rambles across his "farm" in *A Sand County Almanac*, in which the familiar woods and fields come into sharp focus through repeated close attention, and the boundaries of his property come to seem porous and distant. Then there's the conservationist Michael Fay's story (written up as an as-told-to account by the esteemed nature writer David Quammen) of bushwhacking through the wildest tropical ecosystems of Gabon and surveying "every pile of dung, every monkey or antelope or mammal that we saw, every tree by species and diameter within ten meters of the trail," all while keeping a regular journal, an expedition he comes to call the "megatransect." Both of these nature walks meld the narrative of movement to the process of discovery and classification, of finding and identifying and cataloguing and ordering and organizing. Knowing where everything fits.

We can go deeper. There's John Muir, adding an existential dimension to this journey—who can forget him clambering up a tree in the midst of a High Sierra ramble to experience the drama of a windstorm? And not far behind is Thoreau, whose famous essay "Walking" renders the act of tromping around in the woods as a kind of pilgrimage: "We should go forth on the shortest walk, perchance, in the spirit of undying adventure, never to return, prepared to send back our embalmed hearts only as relics to our desolate kingdoms." Versions of this exhortation reappear frequently at the outset of walking narratives—you'll find it, for instance, in the account of Fay's megatransect.

We won't be going that far—no final goodbyes to humankind needed for this journey. Instead, we'll be walking between two landscapes, where biomimetic creatures abound. Six legs, four legs, two legs. Wings and parachutes and the slime trails of gastropods...How ubiquitous the need for movement is among the living; think of a virus spreading, seeds dispersing, spawners searching for mates, predators in pursuit, and prey in flight. It's this diversity that makes the ecosystem metaphor appealing; the real world is full of successful models, tried and tested by natural selection, constantly evolving.

And yet, how easy it is to take something so ubiquitous for granted, to assume that if simple organisms can do it, it must be a simple matter. In the mid-1980s, with artificial intelligence research struggling to make any progress, a cohort of roboticists challenged the assumption that the human mind is where the action must be. The MIT researcher Rodney Brooks offered something of a credo in a brief but provocative critique titled "Elephants Don't Play Chess." The "essence of being and reacting," he argued, lies in

> the ability to move around in a dynamic environment, sensing the surroundings to a degree sufficient to achieve the necessary maintenance of life and reproduction. This part of intelligence is where evolution has concentrated its time—it is much harder. This is the physically grounded part of animal systems.

Movement is more fundamental than mind. He wasn't alone with this "physical grounding hypothesis"—the Carnegie Mellon professor Hans Moravec was thinking along the same lines. "It is comparatively easy to make computers exhibit adult level performance on intelligence tests or playing checkers," Moravec argued, "and difficult or impossible to give them the skills of a one-year-old when it comes to perception and mobility." Before you start developing artificial neural networks, they argued, you should understand what it takes just to *be* in the actual landscape in which living things evolved and refine your definition of intelligence accordingly. More recently, Karen Barad makes a similar point: "We do not obtain knowledge by standing outside of the world; we know because 'we' are *of* the world."

What's important for our journey is the extent to which biomimetics is built into this argument. Animal systems undergoing evolution in the real world—that's the reference point for how to develop robotic systems. Among the robots that Brooks developed as proof-of-concept for the control system he called "subsumption architecture" was Squirt, a tiny creature with a limited suite of sensors and a wheeled platform for moving around. The subsumption architecture provided hierarchical layers of reactive behavior, allowing Squirt to scoot to a dark corner to hide from noises, then roll out into the room when the noise had subsided for a certain interval. Brooks's point was that this quintessentially cockroach-like behavior could be reproduced by a control system requiring just a few lines of code to manipulate sensors and wheels. What seems like the result of a complex decision-making process—*uh oh, someone's coming, the humans are here, run away!*—could actually be produced by the sensors and actuators responding to a simple hierarchical sequence of commands.

Imagine proceeding further with the biomimetic implications of this "physical grounding hypothesis" to include all the different variables that add up to a horse being able to make its way through a show jumping course, or a human accomplishing the morning routine of rolling out of bed and strolling to the kitchen to put the kettle on. A physical body like ours can move in multiple combinations of directions and orientations, known more technically as six degrees of freedom. Three might take us in distinct linear directions; we can *sway*, heading slightly to the left, then correcting and heading to the right as we move forward, or we can *heave* up and down, or *surge,* pushing backward or forward. Our bodies can also twist, *rolling* around one axis, *pitching* or *yawing* around other axes as we rotate through space.

To partake of these possibilities by going for a walk, we need physical mechanisms for enabling movement. That may seem self-evident—we have legs! But as Brooks and his cohort suggested, trying to get robots to move as effortlessly as we do after thousands of generations of natural selection is no easy task. One place to start is the physical mechanisms of locomotion, specifically the muscles that contract and release to generate both forward propulsion and the rotational forces, or torque, that allow us to imperceptibly adjust angles and stay upright as we alternate between one limb and the other. Those muscles often work antagonistically—they create torque by flexing against each other, and they need to be attached to an articulated frame that combines structural support and the flexibility to rotate and bend. That would be our bones and joints, connected via tendons and ligaments. Not every joint offers the same degree of flexibility—the hip joint, in humans, is a ball-and-socket mechanism that offers three degrees of freedom, while the knee is basically a hinge offering just one degree of freedom. In robotics, these physical components are known as effectors and actuators—they're the parts that do the physical work.

In Squirt, the propulsion system consisted of what are now commonplace toy components, just a tiny electrical motor and a set of wheels. That was a start, but those alone didn't mean Squirt would have any way of knowing if it was headbutting a wall repeatedly, or if there was an inviting dark corner somewhere nearby. The same is true for a galloping horse, and for us: we need our senses to gather this information about what's happening around us, and to us, in order to respond. To maintain our equilibrium, we need to be able to sense our position relative to the gravitational pull of the planet, and we also need some way of sensing how we're dealing with those six degrees of freedom, whether we're accelerating through space or leaning to the side or tipping over backwards. In humans, these functions are performed by tiny whorls and membranes deep inside each of our ears, which act like the fluid-filled spirit level we might use to make sure the studs and rafters are plumb and square in some DIY home construction project.

We also need a way of either seeing what's underfoot or feeling it through the sensitive network of nerves in the ball of our foot and the tips of our toes. These are all biotic versions of what roboticists call sensors—Squirt, for example, had three, an infrared sensor for detecting obstacles, a microphone for hearing, and a visual sensor to alert it to changes in light. In our bodies, our sensors work in concert with each other to some degree. Shut your eyes, or enter a dark room, and the sensors in your inner ear will still function, and so will the sensors in your toes, but you'll probably notice that moving feels just a touch unsteady without the usual visual cues.

The nervous system connects all the dots for us. We need systems that can process the information that's coming in and tell our limbs how to respond. In our bodies, the networking runs down our spine from our labyrinthine central processor and branches out into the most delicate electrical filigree, allowing us to sense the hovering of a mosquito wing against the hairs on our arm, and flinch in reply. If we find ourselves in a suddenly dark room, we might need to react instantly, with our reflexes, to avoid falling flat on our face, or we might decide, if we have time, to slow our pace and feel our way forward, planning a route to the door. But on an even simpler level, we need something to set a goal for us, to get over our inertia and decide to make a move, and to do so with the pattern of limb movements we call walking. We need what's known in robotics as a control system, which includes both the capacity to store and process information, and the ability to coordinate our sensors and actuators. Squirt had just a miniscule eight-bit chip for a processor, and the subsumption architecture provided a limited set of rules to follow as data entered and commands exited the system.

There's something else, too, something critically important to the calculus of mobility but easily overlooked. To move, to exert a force that exceeds the inertia brought about by gravity, you need a power source. Energy has to be coming from somewhere, whether it's a granola bar or sunlight or an onboard battery, as was the case with Squirt, and this energy has to be converted into a form that all the systems can use on demand. How fast you go, and how far, are ultimately questions of how much energy you can access, how quickly you can deliver it to where the action is, and how wasteful your actions are. When you study locomotion in animals, you soon realize that evolution tends to favor efficiency. The more "parsimonious," or thrifty, you are with fuel, the more likely you are to reproduce. To make it out of the lab, robotic creatures need to reckon with the question of efficiency as well.

~

The physical grounding hypothesis, it seems to me, is really invoking an existential question—what does it mean to be? What is being in living and nonliving things? That kind of question is the focus of the philosophical discipline known as ontology, the study of being. And Brooks's reply, it seems, has to do with moving and sensing specifically, but more broadly with the mechanics of existence. The physical grounding hypothesis draws upon the evolutionary experience of the living, but he ultimately makes no distinction between the moving and sensing of a cockroach and the moving and sensing of a roachbot. Being, by implication, is not the exclusive domain of the living; the robot is a thing that can be in the world in the same manner as animals like us.

Recently, another version of this idea has emerged in philosophical circles. Known as object-oriented ontology, it describes a landscape in which there's no hierarchy between the living and nonliving; being is not dependent on being alive. One possible result is a "flat" ontology, a "democracy" of things in which there's ample diversity, but no special status for the thoughtful human over the shovel. Instead all entities, living and nonliving alike, are ultimately different conglomerations of chemical elements engaging in various interactions, all of us, to put it in the poet Gary Snyder's terms, "orderings of impermanence." The things on an object-oriented list, however, are not just the traditional hallmarks of wilderness, untouched by human design, but all of our re-orderings of matter as well: vehicles and sofas, laptops, and wilderness areas. These lists are intentionally eclectic, a revisionist discombobulation of traditional hierarchies meant to suggest equal standing for cart and horse—each, they suggest, is worthy of attention. Agency and autonomy might just as well be properties of the nonliving. We're all objects, and we're all beings, or more concisely, "tool-beings," as the philosopher Graham Harman would have it, whose existence and interactions deserve attention.

Biomimetics is a peculiar rejoinder to these subversively egalitarian lists, lists whose central irony is their lack of taxonomic coherence, just a compilation of tangible artifacts and intangible ideas that suggests the limits of human knowledge and authority. What we find in biomimetic research are two categories, living and nonliving, whose boundaries become less stable through the process of trying to find affinities between the two. The metaphoric mirroring that's going on is a kind of embodied translation—by reproducing the mechanism of legged locomotion, we are trying to imagine ourselves into the material existence of a nonliving object, and more fully inhabit what it's like to be a horse or a dog on the move. One of the ways we can understand what it's like to be a machine, in other words, is by understanding what it's like to be a living thing—the mechanisms that make us go, our own biotic thingness. And, conversely, we understand living systems by modeling, or reverse engineering, with machine systems.

What happens, for example, when we start to think more broadly and inclusively about the definition of being? For example, when we begin to rethink what it means to be a plant, a whole series of previously unasked questions emerge about the nature of this suddenly liminal being, this vegetal creature we've long treated as lacking in subjectivity: no consciousness, no intelligence, no agency, no affect. We go deeper into plant being by trying to develop models or metaphors that seek answers to those questions, even if they are ultimately imperfect representations.

There's some solace for the naturalist in this attention to the nature of things. A quintessential nature walk is about noticing everything; it rewards breadth of curiosity and welcomes serendipity. I like to pursue questions that take in everything—I like accumulating mental lists of plant communities as I cross from one landscape to another, keeping track of what the sun and the sky are doing, accumulating leaf patterns and bark textures and the arrangements of seeds, and then seeing what kinds of fungi are pushing through the duff, and adding that to what kinds of birds are singing, and whose dung is so prominently displayed on a trailside stone (and what kind of stone is it?). Think of avid birders, pursuing the unchecked entries on their life lists, or Snyder annotating, in the manner of a list poem, the fecundity of the natural world, the "ineradicable populations of fungi, moss, mold, yeasts, and such that surround and inhabit us… deer mice on the back porch, deer bounding across the freeway, pigeons in the park, spiders in the corners…"

~

Planes, trains, and automobiles: after dealing a blow to the bamboo's expansionary plans, my family and I headed off to visit the in-laws in the Sussex countryside outside London, via routes I had traveled frequently before without the slightest thought to the mechanics of my movement. I wasn't trained in thinking about how I move, but the metaphoric journey had sharpened my attention. From the time we started watching the little white plane icon hover over Boston to the time my family and I exited customs at Gatwick and dragged ourselves onto the tram between terminals and finally lugged our sleepless selves into a taxi, all sorts of wheels had been set in motion to move us along, all of them burning fossil fuel and churning out exhaust.

A week later, I reversed the sequence, off on my own motorized journey to visit the Dynamic Legged Systems Lab at the Italian Institute of Technology in Genoa. On the way across London to the airport, I was half tempted to track down one of those electric scooters that get left around town, partly for the sake of novelty, and partly because I was looking for a jolt. Every form of mechanical conveyance I'd sampled so far seemed to offer such a smooth ride. Did I have to worry about the effectors or actuators? Nope. Sensors? Nope. Control system? Negative. Power supply? I wasn't even sure what the trains ran on anymore. So easy and so fast. Just sit back and let the wheels turn.

Hurtling down the autostrada from Milan in my rented Fiat Panda, I passed signs, in Italian, that appeared to be warnings to slow down during fog, but otherwise it was hard to know why you'd want an alternative to the automobile. I was trundling along, doing the posted 130 kilometers per hour (I had the feeling that was somewhere around 80 miles per hour), when suddenly there was a presence, a shadow over my shoulder, like a lioness bursting from the grass at the rear flank of an unsuspecting zebra. A Ferrari, a neon green apotheosis of everything automotive, hovered for a microsecond or two, then sliced the space at my side. Gone. I didn't even have time to twitch at the wheel, let alone change lanes.

Did I just imagine that?

Then I arrived in Genoa, a transportation hub defined by movement, by container ships pulling in, long haul trucks and railroad cars pulling out…and immediately got lost. I'd booked a hotel in the medieval quarter of the city, the largest in Europe, which straddles a range of dramatic hills above the bustling wharves and shipyards of the Mediterranean's largest port. Not a good move for the automobile-oriented.

I couldn't find my hotel. I literally could not find it, not with all the steep one-way hill climbs and steeper hill starts (did I mention the Panda came equipped with standard transmission?). I was squeezed between balconies, dodging scooters and pedestrians with signs I could barely read warning against parking, stopping, and 20 other possible infractions. It was all historic and gritty and gorgeously scenic, exactly what you'd expect from a postcard—if I could've just paused to take in the view or figure out where I was in this medieval maze, this pile of transportation spaghetti.

After an hour or so, my phone died, and with it my only map. I thought at first I'd roll with it. This was like entering the mindset of the ancient inhabitants—they wouldn't have had a gadget to tell them where to go. I'd just deploy my intrinsic sense of navigation.

I curled around. Down to the harbor and along the waterfront. I passed couples out enjoying the *passeggiata,* the evening walk, with their dogs in tow.

I passed the same people once. Twice. Then again, ducking into a restaurant. Smiling. Laughing. How I longed to join them. Just ditch this wheeled vehicle and set out on my own two feet.

Late afternoon stretched into evening. I felt as if I was operating not just a different mode of transportation, but in a different dimension. They were living their lives at one pace, experiencing one version of time and distance, and I was living mine at another. I could go a mile in a hurry, but at the same time, I couldn't get anywhere; I was like a hamster on a wheel, looking out through the glass of its cage. I was caught in what felt like the same moment, endlessly repeated. Think of the inefficiency. The fuel I was burning. The climate I was changing. What was I doing?

Being directionless didn't put me in touch with the ancients either, unfortunately. My alienation was more structural; the city itself wasn't built for the likes of me. When these streets were brimming with medieval citizenry, they were all marching to a different beat. They wouldn't have been caught nose first on a vertical one-way street that dead-ended in a lane barely wide enough for a donkey cart to squeeze through, trying to push the clutch into reverse and whimpering plaintively, like a terrified kitten.

I know what you're thinking. *Why didn't you just...* Yes, indeed, why didn't I? All sorts of solutions occur in retrospect, but in the moment...no. Remember how easy and fast it was to go down the road? The flip side is it's not so easy to deviate, to make up a Plan B on the fly, to improvise. Roads and cars are not meant for improvisation, even in Italy. They're meant for you to follow the signs and go with the flow. How do I know? Because, as you can imagine, I left a lot of shaking fists and livid faces in my wake, accompanied by a torrent of advice that got more emphatic every time I hesitated. Like the moment I found myself facing the end of the road at the top of a hill—for some reason known only to the ancients, the way suddenly went from auto friendly to alleyway. I was going to have back down to turn around without crunching somebody's parked car, and sure enough, as I looked in the rearview, a face appeared. An exasperated face, in a helmet, belonging to a scooter operator. And behind her another. And behind him another. Needless to say, the gang did not take kindly to having to roll backwards down the hill with me.

To break free of this nightmare, I drove out of the city, found a shopping mall on the outskirts that boasted an actual parking garage, and went inside to the food court. Of all the places in Italy, here I was in a mall food court. Why? because I could park and get out of the car. There was still nowhere to charge my phone, but at least I could use the Wi-Fi to check the map on my laptop. Yes, that pin on the map was the hotel, that was clear as day, and there was the sea, and...there was the zigzag maze of circuitous, one-way streets surrounding everything. The blue "suggested route" looked like someone had been practicing their signature in cursive, something with lots of loops. When I closed the laptop, what stayed with me was not a sense of where to go, but a growing sense of hopelessness and panic.

Back in the car again. No Wi-Fi. No mobile data. None of that 21st century navigation stuff.

Darkness had arrived and the streetlights, few and far between as they were, winked on beneath the orange haze of sundown. It was all so lovely—what a beautiful, nerve-wracking, utterly incomprehensible city. I risked stopping with the flashers in the middle of the street and accosted two women chatting over their leashed dogs. Did they know where the Best Western was?

They waved vaguely, their hands motioning like they were throwing a crumpled ball of paper at a wastebasket without much hope of making it. One waved up the hill. The other waved in the other direction, down the hill. Maybe they were both right.

By ten at night, I was on the verge of a nervous breakdown. I'd paid for a room. But I couldn't find it. And I was stuck with this car. Literally stuck in it. No place to legally park it. It was like I'd been chained to four wheels. What was I thinking? This was a book about walking—I should've taken that literally and walked here. Walked the whole way. I'd rather have walked from Sussex rather than endure this madness. Swam the English Channel, sauntered down the Riviera, the whole thing. As long as I never had to get behind this wheel again.

The city was wrapping up for the night—the last businesses were closing and I was still roaming the streets. I imagined myself at 3 AM, running low on gas. I imagined dawn breaking beautifully over the shimmering Mediterranean, with me nodding off at the wheel. I had to stifle a strange feeling, what might have been an actual sob coming up from my chest.

What the hell was I going to do?

There was a Novotel down by the waterfront where the cruise ships dock, on the other side of an overpass and a roundabout. It was too frantic with traffic to stop earlier, but now I could pull to the curb by the front doors and leave the flashers on.

Look, I don't need a room, I confessed to the front desk clerk. *I have a room already booked someplace else. But the truth is, I can't find it. I've been driving around for hours....I'm wondering if I could possibly just leave my car here overnight. Just leave it here with you, basically, and just walk away....*

They were kind to this desperate traveler. In under ten minutes the Panda was parked underground and I was in a taxi, racing back into the medieval quarter. I knew all the streets by now. Except...what was that turn; I never saw that. And then, okay, there was the opera house, but which way was he going now...

In all those hours of self-navigation, I never once passed this way. The final insult: my lodging was set back off a winding cobbled side street, in a pedestrian-only courtyard.

No cars allowed.

Next morning, I took a serene cab ride through truck clogged tunnels and scooter crazed viaducts up into the hilly suburb of Morego, where IIT sat in a rectangular bunker of a building cut into the side of a steep grade. After filling out some paperwork, I passed through a set of airport-worthy locked glass doors while a call was made to the home of HyQ, the quadruped robot I'd come to see.

Claudio Semini, who heads the Dynamic Legged Systems Lab, escorted me back through the lab space that several teams share. He has the compact athletic build of a football striker, and he moved quietly but briskly through workstations that were crowded but nearly silent except for the tapping of keyboards. He's spent much of his career here, arriving in Genoa in 2007, two years after this branch of IIT opened its doors, after a stint in Japan working on robotic limbs under the influential roboticist Shigeo Hirose. As one of the first cohort of PhD students at the Genoa campus, he built the prototype of HyQ as his dissertation project.

The rooms were long and divided lengthwise by a series of glass panels. On one side, a cramped line of desks sat beneath shelving that reached the ceiling, cluttered with the mechanical odds and ends of robotic construction. Graduate students and postdocs sat staring intently at their computer screens; a few huddled in clusters around a monitor. What immediately drew my attention, however, were the locomotion experiments on the other side of the glass walls. Each nearly empty chamber held a work-in-progress, a creature awaiting animation. We passed something with long prehensile arms mounted on a wheeled base—a prototype version of a creature known as Centauro, a humanoid set of prehensile limbs attached to four legs that ended in wheels, like roller skates. We passed a humanoid in harness, suspended from the ceiling over a repurposed treadmill. The glass was necessary to isolate experiments from background noise, but it all felt, in an eerie way, like a zoo, or the cages in which research animals are held. I had the illogical but nevertheless unmistakable sense that one of these things might get off its leash and get out.

Semini and I sat down in front a computer monitor, and he pulled up some PowerPoint slides. It's easier to describe the technical details of quadruped gait dynamics—the way a horse or a dog changes the cadence of its footfalls as it shifts from a walk to a trot to a sprint—if you can see the diagrams. He showed me how a quadruped animal's legs form a rhythmic elliptical pattern as they move forward, the shape of which changes as the animal picks up speed or slows down. These patterns were not discovered by Semini himself; he's an engineer by training, and he's not going to be out on the African savannah measuring all the different torques and velocities that a cheetah generates in pursuit of an antelope. That work is usually done by biologists, who gather the data and transform their observations into mathematical models, the kind of information that an engineer like Semini can translate into robotic form.

The evolution in the physical shape of robots and their capabilities partly depends, then, on the pace of innovation in biomechanical research. In the case of quadrupeds, that research has a long history. Some of the earliest research had to do, not surprisingly, with horses, since these research subjects were already trained to trot, canter, or walk on command. In 1878, the pioneering photographer and inventor Eadweard Muybridge performed a revelatory analysis of horse locomotion by arranging 12 cameras along a track and capturing a sequence of photos that showed, for the first time, what was not apparent to the naked eye: all four of the horse's feet left the ground at certain intervals, folded toward each other in a rough triangle. Muybridge went on to use the same stop motion technique with a veritable menagerie of creatures, including baboons, elephants, ostriches, and camels, often with the crosshatching of a zoo's cage in the background. To see the pattern of footfalls, to find the beat in the thundering of hooves, we needed a method for slowing down our experience of time so our senses could catch up, which the cameras provided. It was an early indication that we are surrounded by movement, immersed in it, and yet much of it is taking place at scales either too slow, or too fast, for us to notice.

Semini mentioned an influential study by Richard Taylor and Donald Hoyt, published in the journal *Nature* in 1981, in which miniature horses were trained to run on treadmills and fitted with equipment that measured their oxygen consumption at different speeds, marking particularly those moments when the animals switched gaits to maximize efficiency. We humans do the same, Semini said, with our two gaits: walking and running. At the gym, we might set the initial speed on the treadmill so we're walking, enjoying a relatively efficient gait, but then as we kick it up a notch we reach a point where we can sense that the amount of oxygen we're gathering with every panting breath doesn't match the progress we're making with our pace, and it just makes intuitive physiological sense for our body to break into a jog. Horses demonstrate a similar calculus. Each gait has an optimum speed in terms of energy expenditure—walking too quickly or trotting too slow are both relatively inefficient when compared to shifting to a different pattern.

Just as Muybridge deployed what was a novel procedure at the time—stop motion—to get a more accurate portrait of gait dynamics, so biomechanical research today is taking advantage of technological advances to gather data in the wild, on the rough terrain that an animal actually has to negotiate instead of the flat and confined artificial spaces of the past. Instead of observing cheetahs chasing a furry toy in a zoo, as was previously the case, researchers now work on the cheetah's turf in Botswana, using specially developed collars in which a suite of miniaturized sensors measure speed and acceleration and even the twists and turns taken when these predators are in pursuit of big game. All of this data would be of interest to biologists, who now have a more detailed portrait of what it means to be a cheetah, or more broadly, what "being in the world" means, if we recall Brooks's definition of the essence of being as moving around and sensing the surroundings. And all of these torques and angles and gait patterns—all of this more granular, more precise data—would be useful to those trying to fine tune their robotic representations of moving around in the world.

For Semini, early inspiration came from perusing old dog breeding catalogues from the 1950s, which demonstrated the range of morphological possibilities within just one species. There was the squat and broad-shouldered bulldog, and nearby was its antithesis, the fleet and dainty whippet. From a locomotion point-of-view, each was a different potential platform. Even an untrained eye can spot the difference between a sprinting greyhound and a lumbering Rottweiler, but on a deeper level, these physical differences change the calculus of locomotion, impacting the energy cost of using one kind of gait before reaching a threshold and switching to another. One canine version might offer better top speed; the other might deliver more robustness, be more capable of withstanding the body blows of a hostile world.

Semini's initial studies focused on the hip, and the way the action of that joint influences all the other variables of balance and foot placement and coordination between limbs. He began with a mechanical engineering approach, in other words, one that reflects Brooks's emphasis on getting the "physical grounding" right first. In a photo from his graduate school days, Semini poses next to his prototype, whose name is short for "*hy*draulically-powered *q*uadruped." He's crouching, so that the three-foot tall creature seems almost level with his shoulder as it leans back on its bent hind legs in classic canine repose. It looks like a sculpture of an animal, fashioned from gleaming alloys, coiled in tubes that dangle, umbilicus-like, from the ceiling. More striking, however, is what was missing from HyQ's early portrait: in keeping with the focus on joints and the mechanics of motion, the robot appears to be all torso and legs. There's a bundle of hydraulic tubes emerging from the front of the body cavity, similar to the power brake lines through which fluid can flow. But no head.

Semini had to step out for a conference call to update the grant donors on the team's progress, and Victor Barasuol, a postdoc who specializes in control systems, joined us in one of the glass rooms where an early version of HyQ awaited. Like the others I saw earlier, the robot was suspended about a foot in the air from a harness in the center of the room.

My first impression was of a machine, not an animal. Then, perhaps, an animal in one of those skinless muscular-skeletal diagrams you might study in a veterinary class. Everything was exposed. The "torso" was built around an aluminum rack whose ends curve outward, less like a ribcage than the handles you might find on an ambulance stretcher, while the metallic legs were tipped in black rubberized caps that reminded me of the crutches I once hobbled around on. Power cables, and data cables, and hydraulic tubes dangled from the ceiling like technological lianas, as if the neurological and circulatory systems normally embedded in an animal's body had been extracted for visual effect. They connected to computer stations at the side and a monitor overhead in the corner, where the simulation of the robot's movement could be displayed as a virtual counterpoint to its actual progress. Before he departed, Semini plugged in a motorized pulley that lowered HyQ slowly to the surface of a treadmill, and the two men guided the robot's 200-pound body to a landing in front of a scattering of bricks.

Clamped to the front of the formerly headless HyQ, at the end of a post whose curve vaguely resembled the neck of a black swan, was a box that jutted forward and pointed slightly down. It looked, not surprisingly, like a head, and it functioned a bit like one, since this visual sensor allows the robot to gather detailed information about what lies on the path ahead.

Its presence reflects the rapid evolution in the robot's control system. HyQ was indeed "blind" in its early headless phase, intentionally deprived of any visual information gathering apparatus so the research team could focus on its reactive capabilities. It possessed sensors at key points on its body, especially the joints, so it could gather data about its orientation and movement, and it's processing capabilities included a "control pattern generator," a software program designed to manage the cadence of putting one foot in front of the other. But because it couldn't "see," it had to operate with no visual orientation and no map, and thus no capacity to guide its steps.

The earliest HyQ, like a visually impaired Squirt, sensed its surroundings and reacted with movement. It followed a pre-installed gait pattern, going forward and reacting to nonvisual sensory data as it planted each step, as we might if we found ourselves in a pitch black room; that's one of the reasons we have so many tactile nerve endings in our feet, so we can keep abreast of what's going on down there and use our reflexes to respond to the stimuli we're receiving, even if we can't see what we're stepping on. I recently experienced this firsthand—I was standing in our driveway, pulling weeds from a raised bed, when it felt as if the ground beneath my right foot was shifting and undulating, almost like beach sand when a wave retreats. I reacted automatically by lifting my foot even before I looked down to see a black rat snake, about four feet long. I'd been standing on its neck. My next move was reflexive, and it involved verticality.

Our reflexes are always ready to be triggered if we stumble, whereas periodic gaits involve volition: we choose not just the speed and cadence of our travel, but also select one path over another, optimizing the placement of our feet to avoid rabbit holes or awkward stones on the road ahead. Usually, we shift between the two, but not equally; we normally have other things on our mind than forecasting our next footfall. Barasuol mentioned stairs—we don't look at the stairs while we're climbing them, deciding where to plant each step. We're not conscious of what our feet are doing until we trip, and then the reactive mode takes over.

To improve, HyQ had to stumble. The team put various obstacles in its path, then watched how it handled the unsteady terrain. Which explained the bricks, and a few other things as well, like the big red emergency button on the side of the torso. At first, I was waiting for someone to flip the switch so HyQ could flex its hydraulic muscles and break into a trot, but I soon realized that wasn't going to happen. Semini and Barasuol bent down and pushed the robot along to demonstrate how it would step over bricks in its path, but it was clear that animating 200 pounds of extremely expensive, bespoke engineering would be a bit like launching the space shuttle; you have all of these different systems that need to function properly, software, hardware, valves properly connected, commands rightly executed, sensors exactly calibrated, or things may go haywire. Hence the red button, and the catcher's mask over the visual sensor, and the bunched towel padding the front post of the frame that held HyQ aloft, all in case the robot stumbled and careened off course, or whoever was running the treadmill panicked and hit the stop button, as Barasuol said sometimes happens.

At the same time, making things tough for HyQ—pushing the machine to the breaking point—was kind of the point of this research. "You need to have in mind the rough terrain from the start," Semini told me. "Otherwise, what's the point? It's just an academic exercise." Robustness is the key to being useful outside the lab. As I well know, if you're rolling down the flat and straight autostrada without a care in the world, four wheels are great. It's only when you get off the beaten track that the wheel stops making sense and the leg becomes an intriguing option. These creatures might be used to clamber over the debris left in the wake of a disaster like the stricken Fukushima nuclear plant, where the radioactive conditions are too unsafe for humans or rescue dogs to explore. Undaunted by the risk of tripping and crashing into a pile of rubble, HyQ could ascend stairwells that would be insurmountable for a robot mounted on a wheelbase.

With conditions like these in mind, bricks and packing crates and wobbly lengths of plywood were only the beginning of HyQ's trials. Video published by the lab shows HyQ in reactive mode, trotting along an ersatz riverbed of round stones and sideways planks laid out on the concrete floor of a warehouse—some of the rocks skitter out from beneath its feet, but the robot's sensors adjust the stiffness in the joints, damping it down to allow more shock absorbency, kind of like the front shock on a mountain bike. It's a strange tableau—there's a handler jogging along beside the robot, holding a loose leash, not unlike the trainers you might see at the Westminster Dog Show trotting their prize canines around.

At one point, the researchers wondered what might happen if someone tackled the robot, so they installed a 50-pound punching bag next to the treadmill, which they swung around to strike the robot in the torso to simulate a kick. Would HyQ be able to stay upright? The blow struck at different moments in the gait pattern, sometimes with a leg raised, sometimes with the feet grounded. In videos of this experiment, the robot is trotting along innocently enough on its treadmill when a nearby human releases the bag. The robot can't see it coming and is knocked sideways, but then it steadies itself by readjusting its outside feet placement to absorb the blow and remain upright. It keeps going.

That's how a dog would handle it, you find yourself thinking. A transformation takes place as soon as HyQ begins to move. It's no longer just a mechanical platform. Despite the cane-like clacking of its feet and the pneumatic whooshing of its "muscles," the way it moves evokes familiar associations: there's a deer poised in the middle of a creek, stepping slowly away with its tail raised in alarm; there's a puppy leaping up to play, bracing itself when you push it away, then vaulting back toward you. There's something in the way HyQ makes its way over stones and bricks that lends itself to imagining it as an urban animal, trotting up the steep and uneven cobblestones, heading right past a driver as they sit imprisoned in their Panda.

A lot of this recorded movement was taking place blind, strictly in reactive mode, but that wouldn't ultimately mimic the way an animal negotiates a difficult route. Animals like us toggle back and forth between reacting and decision-making. A dog loping along a scree slope might be in reactive mode while it sniffs the air for news, but from time to time it might decide to shift to planning mode, making choices about which route to pick through the rubble.

The next step in HyQ's evolution was to provide the ability to see and plan a route. The team proceeded in stages, adding downward-facing cameras first to allow the robot to respond to visual data when it placed its feet instead of just reacting blindly. Eventually, they positioned two cameras in the black box at the end of the swan-like neck, which the robot could tilt up and down autonomously so it could see what tribulations lay just ahead and further along. Lower down, they added a stereo vision camera, which bounces pulses of laser light off the ground, allowing the robot to generate a detailed three-dimensional map of the immediate vicinity. Video footage of these experiments unfolds on a split screen: in the upper left is the camera view, showing us the robot's point-of-view as a chunk of wobbly plywood looms ahead, while the main screen shows us a conventional sideview of the robot making its moves. We can, in effect, look simultaneously through the eyes of the robot and watch it from the outside, another manifestation of the way this technology renders the world both familiar and alien at the same time.

Semini returned after the call, and we headed to another small room that functions as a workshop. Imagine a shop where bicycles are repaired but also built from the frame up, and you'll have a good idea of the array of tools and parts covering every surface. Or maybe it would be more accurate to think of it as the workshop where custom supercars are built by skilled artisans. Semini pointed to a shelf high on the wall, to what looked like the pneumatic cylinder and rod apparatus that closes a storm door. It was, in fact, a bit of evolutionary history, a technological fossil. This was the original prototype for HyQ's limbs, in which the hydraulic cylinder slotted into the rear of the articulated leg.

On one repair stand perched the disarticulated carapace of HyQ's successor, HyQ2 Max, with insides encased in aircraft grade alloy painted black, two burly collars extending fore and aft and hip joints protected by round yellow discs, all with an eye to making the robot more robust for outdoor action. It looked like a giant beetle, or a rottweiler.

But what lay ahead was even more rugged, and more refined. Semini had formed a partnership with an auto parts company called MOOG—they make components for steering and suspension that you can buy off the shelf, but they also design custom parts for Formula One racing. "Let me show you the latest," Semini said, with an unmistakable air of accomplishment, as he held up what looked like an exquisite titanium sculpture of the joint in an animal's leg, perhaps the hock (rear ankle) of a greyhound in which you can see the slender knob of bone with the tendon curved snugly behind it, and the blood vessels that bulge there. All the cumbersome components that burdened the early HyQ would be integrated into channels inside this joint—there are channels for the hydraulic fluid, and others for the spectrum of wires and cables. Actuators this light and efficient could be paired with the battery-operated hydraulic system MOOG was developing, allowing this new canine robot to run around in the great outdoors without being tied to external compressors.

I hefted the new component in my palm; it was surprisingly light for something so solid.

As I tried to imagine it connected to a robot traipsing through the medieval quarter, I was reminded of something Barasuol had said earlier, when we were looking at the prior HyQ dangling over a treadmill. "We don't want to adapt our environment to the machines," he'd said. "We want the robots to adapt to us." Advances in mobility aren't all about moving creatures; to go places, you have to go through places—you have to move through the environment, and that isn't just a set of physical constraints like gravity, it's an ecosystem full of other creatures, the natural histories they've accumulated and habitats they've built and adapted to each other. What Barasuol was saying suggests we can push the physical grounding hypothesis further: we're part of the environment in which a robot moves around, senses, and reacts. The physical world is not just a quarry for ideas, or a testing ground; it's also the place we inhabit, and there are planetary consequences to how we approach the question of efficient mobility.

Just as we might describe a landscape of fear, or a landscape shaped by risk assessment, so we might also think of landscape as a reflection of our technology. The technological landscape is the built landscape of crisscrossing and highways and the exhaust simmering above them, the systems we've constructed around this single answer to the question of mobility. My fiasco from the night before offered ample proof for what happens when the technology can't adapt to us, to our history and habitat. The car says suburb and freeway and strip mall parking lot; an ancient city like Genoa speaks another language, and the resulting negotiations are awkward at best. Biomimetics, with its constant recognition that the mirror is hazy and incomplete, is a reflection of that same awkwardness, and an attempt to find greater clarity.

At the same time, what does it mean for a robot to adapt to us? Are we looking for technology to fit into our world as it is, or to help us transcend its challenges? To me, it seems likely to be both. We want technology that respects who we are and who we have been, but at the same time helps us confront problems that can't be solved otherwise, that helps us experience new ways of what it means to be human. We're adaptable—that seems to be one of the hallmarks of our species. Another seems to be the propensity to build ever more elaborate devices to counter the constraints of our own biology. What biomimetic technology is really doing is laying bare this ongoing negotiation between the preservation of our place within this teeming community of species, and what seems to be our human impulse to create something new, to redefine human existence by dint of our own intellectual prowess.

HyQReal continued to evolve after my visit. It emerged, not onto the hilly medieval streets of Genoa, although in theory it could trot around the cobblestones quite nimbly. Instead, it appeared at a nearby airport hangar. In a video that began with all the pulsating musical and visual flourishes of a *Top Gear* montage, HyQReal looked robust yet sleek, far more like an animal, like a creature ready to be out in our world, than the HyQ I'd seen harnessed in the lab. The difference reminded me of the physical grounding hypothesis, and how much progress was apparent here in the evolution of sensing and reacting to the real world.

In the action sequence, the robot was attached by a cable to the front wheels of a small airplane that still weighed over three tons. As the engineers gathered to watch, it braced itself into the load and began to tug, straining against the cable with a jerk and then rocking back slightly on its hindquarters as each small step gained a purchase. As the forces changed between each hard-earned step, you could almost feel the robot sway, nearly losing its balance, but it compensated for these sideward shifts in momentum by rotating each of its legs and bending at the hip. It wasn't quite reminiscent of a dog pulling a sled across the pack ice, or a horse hauling a buggy along a road. There were elements of both, I'd say, but it was also unlike those biotic counterparts. It had its own way of moving.

The plane rolled slowly forward. I looked for the joint I'd held in my hand, imagining the pressurized fluids coursing through the lines and valves to power the hydraulic actuators.

But it was hidden behind the body's armor. As one would expect in the great outdoors.

~

Two weeks after my drive around Genoa, I journeyed to another branch of IIT, this one located on the outskirts of Pisa, and reached my destination without incident. Maybe because I'd decided to heed the warning of my daughter, Beatrice: "Dad, you're not driving anywhere are you? Because you know what happened last time!" From the airport there was a shiny new tram that took just a few minutes to arrive at the central train station. And this time, I'd booked a room across the street—a one-minute walk from the plaza. No cars required.

Pisa is another ancient city, but the topography is vastly different. The city is built along the banks of the Arno as it meanders toward its mouth a short distance away, and there are no obvious urban mountains in the vicinity, or even the quintessential hilltop villages for which Tuscany is known. In the morning, I walked straight across the street to the train station and rode 20 minutes out through the broad plains of reeds and grazing cows to Pontedera, where the train stopped in front of a concrete gate with the name Piaggio inscribed upon it. Stretching into the far distance in either direction were the single-story assembly lines that were rebuilt after the factory was bombed to ruins during World War II. What the company is best known for is the Vespa, which means wasp in Italian, those iconic scooters zipping around the streets of Rome with Gregory Peck at the handlebars and Audrey Hepburn in tow. A classic, partly because of what it represents: technology adapting to the contours of city life with a degree of grace. Supposedly, they're still built here, but most of the buildings seemed idle now, and the place had a sun-warmed dusty silence about it, as if the crowds of workers who once emerged from the trains toting lunch pails have dwindled in the age of global supply chains.

You wouldn't necessarily connect this manufacturing center with research on plants—there's not much in the way of vegetation in the immediate vicinity. But one of the former factory lines has been converted into lab space, and it's here, within IIT's Center for Micro-Biorobotics, that the robot known as Plantoid has taken root. Just thinking about the Plantoid robot destabilizes the assumptions I've always made about mobility. When I started contemplating this journey, my mind immediately went to robots modeled after animals with legs. The image I had was of a robot running around—a robot not unlike HyQ. Plant mobility, and all it implies, never occurred to me. I had to reconceptualize what it means to move in the world, to sense and process and react.

Emanuela Del Dottore, a postdoc whose background is in control systems, took me back to the lab while explaining that she often finds herself conducting plant-related research. "We have to talk to people who have different backgrounds," she said with a chuckle, as if remembering some particularly head scratching incident of interdisciplinary miscommunication, "so it can be quite challenging to understand each other. But you also grow as a researcher." That communication across disciplines is characteristic of biomimetic research generally, but one of the key differences between this research and the research on quadrupeds, and perhaps why it seems so groundbreaking to me, is that there's no Muybridge to draw upon; there's nobody who began investigating the phenomenon of plant mobility a century ago. As the center's director, Barbara Mazzolai, told me over the phone before my visit, many of the questions here haven't been thoroughly explored before, because until quite recently we simply didn't imagine the need to ask them. Consider, for example, Hans Moravec describing the difference between plants and animals like us in an influential paper from the early 1980s: "Trees are as successful and dominant in their niche as humans are in theirs, but the life of a tree does not demand high-speed general purpose perception, flexible planning and precisely controlled action." Trees, he argued, evolved along a "sessile" or stationary path which presented a limited set of challenges, while mobile creatures like us had to be quick on our feet to survive all the new tribulations we faced as we roamed from place to place.

It's only when you begin to reverse engineer botanical strategies for survival and reproduction that you begin to recognize the inaccuracy of our previous judgment—plants are not the biotic equivalent of a stationary solar panel, just sitting there in a persistent vegetative state, soaking up the sunshine. Instead, the Plantoid team prefers a new paradigm of plant behavior, articulated in a vision statement they published in the decidedly non-robotics journal *Plant Signaling and Behavior.* Plants, they contend, "are dynamic and highly sensitive organisms," with roots that behave "almost like active animals, performing efficient exploratory movements, with the root apices that drive the root growth in a search for air, nutrients and water to feed the whole plant body."

Like horses and humans and cockroaches, plants need to sample the world via a suite of appropriate senses, and they need to process this incoming information and "decide" how to respond. Thus far, Mazzolai's team has identified several potentially useful "tropisms," mechanisms that plants use to sense the world and direct root growth. Gravity is one—plants can sense the gravitational pull of the Earth. Phototropism, or sensing and responding to light, is another; together they allow plants to "know" which way is up toward the sun and which way is down into the soil. Plants typically respond to these stimuli by secreting two key growth hormones, auxin and cytokinin, that govern cell growth and distribution in the root tip. It turns out, however, that even in the zero gravity conditions of the international space station, 240 miles above the Earth, plants are able to compensate using other mechanisms that are still only partially understood.

Chemotropism, growing in response to a gradient of positively or negatively charged ions in the soil, may be one of these compensatory mechanisms. Roots use soil chemistry like the coordinates on a map, detecting greater concentrations of the nutrients they need, like nitrogen and oxygen, and responding with the hormones that stimulate cell growth on one side of the root, charting a course in slow motion. When a root hits an obstacle—like a rock—it senses this chemical dead end and follows the trail of ions in a detour around it. It can also grow away from ionic concentrations that might threaten its life, by following chemical gradients away from floodwaters, for example.

What's intriguing to me about this new paradigm is the way it expands our sense of what it means to be in the world, while at the same suggesting how different the mechanics of existence could be. While Plantoid research might begin by describing the biomechanics of root growth, just as a quadruped roboticist might consult the gait dynamics of a horse, the mechanism itself bears little resemblance to animal locomotion. As the Plantoid team explained in a subsequent article, movement by root "elongation from the tip" is grounded in the distinctive "architecture" of the root, the physical features that change both regionally and at the cellular level as we approach the end, or apex. The region furthest back from the apex is older and tougher, its mature cells already arranged to form a rigid base for the expansion of the root. Closer in, there's a zone in which newer cells are elongating as they mature—as they absorb water and their membranes stretch, they act like a coiled spring, pushing the root tip from behind. These cells originate in a smaller region nearer the tip; as cells divide and grow larger here, they too exert forward pressure on the root cap. The root cap itself, the pointy tip of the drill, is composed of cells designed to slough off in response to pressure from behind; before they do, however, they secrete a "mucilaginous" lubricant, creating a layer of gooey slime and excised cells to counter the friction of pushing aside soil particles. All of these physical mechanisms add up to energy savings when it comes to squeezing along underground, making apical extension a more efficient mode of transportation than burrowing like a mole or pounding away at the surface like a pneumatic drill.

Del Dottore led me to the first Plantoid model, designed to explore soil penetration, tucked away on a wheeled cart in the corner of a shared office space. Actually, there was a small grove of robots here, all smaller than I expected after the life-sized quadrupeds whose presence seemed to fill the room. These tree-ish figures looked like baobab bonsai—a squat brown plastic trunk, wide at the base, crowned with a few skewer-like limbs, perched on a clear platform. Mazzolai had already warned me that the "trunk" had no functionality at this point. They chose this shape, she said, to reflect "the deep analogy of a biological teacher." She sees the terms biomimetics and bioinspiration used interchangeably in the field, she told me, but the term biomimetic is closer to her original idea—to emulate, as closely as possible, the way plants behave as active beings in the real world.

Beneath the platform with its protruding trunks, a clear-walled box contained several appendages that dangled above a patch of artificial turf, while another "probe" rested in a container of tiny white foam pellets. These were the robotic version of tree roots. They looked like stubby troll fingers wrapped in green twine; three small springs had been attached at what would be the knuckle, so that the cylindrical tips, composed of a soft silicone material, could take different angles as they performed the act of burrowing. The roots looked nothing like HyQ, yet the two robots shared the same three basic elements of bioinspired design: the physical features, or actuators, that provide the power and make the parts move, the sensors that gather data about the environment, and the control system that puts the two together to emulate the behavior of the creature, be it dog or bamboo.

Del Dottore handed me one of the robotic roots, still leashed to its green filament—it was about the size of a soda can but heavier than it looked, densely packed with electrically motorized gearing and the processors running the control system, as well as a flashing green LED light that illuminated the entire soft tip for biotic effect. She explained the original experiment by picking up a handful of artificial soil, in actuality a layer of pea-sized white foam beads that filled one of the boxes. They used this "cohesionless granular medium" for experiments because all its properties are known and can be held constant, whereas actual soil is a kind of stone soup full of all kinds of grated chunks and sticky bits. A video monitor in the corner above us played a time lapse sequence of corn seedlings sprouting white roots that wriggled down into translucent green gelatin, then split and fanned out. To quantify the movements of roots in response to different chemical gradients, the team essentially made lots of batches of Jell-O, each with a different chemical composition, then studded each one with corn kernels whose growth they filmed. It made me think again about Muybridge's parade of animals, all traipsing in slo-mo. Time lapse is the very antithesis of stop motion, and yet both techniques allow a temporal translation to take place, so we can experience what it's like to be a plant, to sense and process and respond to the world at a nonhuman pace. Plant time forces us to consider the nature of subjectivity itself, that vanishing point where subject and object meet.

I rubbed my finger across a tiny rectangular chip embedded on the surface of the silicone, which Del Dottore said was the result of collaborative research with a team in Barcelona. Coated in a special polymer, it was one of several sensors embedded in the soft "apical" region of the robot, including sensors for gravity and temperature as well as a customized touch sensor in the very tip, and more recently, innovative sensors for detecting nitrogen, phosphorous and pH. These sensors allow the robot to react to the nuances of the soil as if it is governed by various tropisms.

Del Dottore uncapped a vial of water and raised it to a round white sensor at the tip of one of the roots. After a delay, during which the processor dealt with the incoming sensory data, a whirring, whining sound of mechanical gears began to emanate from deep within, and the robot slowly but surely tilted sideways, as if its tip was drawn to the water. Inside, the plant-inspired software was directing the gears to tighten one of the three springs that run from an external attachment point down through the interior tip of the robot, shifting the angle of the artificial root tip's descent. It was meant to represent what plants do when they hunt for water or hit a rock—the cells on one side of the growing region continuing to elongate, while those on the obstacle side remain relatively compact until a bend in the root forms.

But there was a further stage of innovation on display here, which was easy to overlook unless you recalled the armored physique and piston-like joints of HyQ. The moving parts of that system were never meant to be malleable; their inspiration might be biotic, but their actualization was clearly something Jeremy Clarkson might adore. Soft robotics, on the other hand, takes a different approach, one inspired by what might allow an octopus to squeeze through a tiny hole in the lid of an aquarium, or reach out and grasp your wrist with its suctioned tentacle: as with a growing root, there's no predefined body.

"If you are rigid," Del Dottore said, patting her fist with the palm of her hand, "you're done. Soft materials, however, you may squeeze and pass through."

She showed me a root that was looped in what looked like the plastic cord we used to braid into boondoggles at summer camp, but which was actually a "thermoplastic polymer" that becomes sticky at 200 degrees Celsius without melting entirely. Using this material as an actuator was inspired by the way roots grow by producing soft cells near the apex but elongating and hardening cell walls further back. As this robotic root forges ahead, following some chemical gradient in the soil, the polymer cord is meant to rub past a heating element inside the cylinder and emerge as a sticky coil that unfurls in tight loops at the back, creating a hollow tube that is flexible enough to bend at first, then hardens as it cools.

In plants, this older, hardened growth provides a physical connection back to the stem and from there all the way up to the leaves and down to the other roots. Hormones, nutrients, water—the plant has no mammalian nervous system, but it can still send and receive chemical signals between its extremities, and it can still make decisions in response. However, this communication mechanism is passive; it happens implicitly, via chemical feedback loops. There's no central decision-making organ sending out orders, which can be a difficult model to describe, even for those who work with it every day.

"Don't call it a brain!" Del Dottore admonished herself with a laugh at one point, when describing the way a plant responds to changes in the soil. "It's intelligence."

Plants have no brain. Yet they have intelligence.

I had to take a moment to wrap my own brain around that one once more. Plants are not conventional subjects. They're not conventional objects. If biomimetic robots are machines that exist in ways we normally associate with the living, then plants are the opposite: they're creatures that live in ways we normally associate with objects. They make the binary between subject and object seem all the more richly convoluted, a space populated by all kinds of possibilities, robotic and otherwise.

Much of the Plantoid team's work, Mazzolai told me, involves measuring and understanding this decentralized decision-making process, this botanical vision of intelligence distributed throughout the system rather than concentrated in leaves, stems, flowers, or roots. "You can see they take a decision because they adapt their bodies," Mazzolai had said. "How do they manage all this complexity?" There's still little knowledge, she'd said, of the way communication happens between various structures within the plant. One of the reasons for developing the soft polymer-based root system with its hollow core was to enable the possibility of running some kind of physical communication system between the branches above and the root tip below, a system that would allow for plant-inspired decision-making to take place throughout the robot, connecting multiple botanically inspired features.

I pointed to what looked like tiny leaves on the plastic branches. They resembled strips of glossy packing tape, coated with a polymer that gave them a pinkish sheen when they caught the light. These were passive actuators, Del Dottore said. Another seeming contradiction. In Genoa, the main challenge with hydraulics had been delivering energy efficiently and powerfully to the actuators; I'd just assumed that active and actuation were synonymous.

A golden sheaf of wheat was sitting nearby, next to a pair of pinecones—they were more than just decoration, apparently. Del Dottore plucked one of the husked kernels, wet her fingertip with water, then stroked the twin wispy spirals that splayed out from the husk like wayward antennae. They reacted by straightening toward each other until they nearly met, a motion that, in combination with drying and separating again, scissors the seed into the soil. Pinecones use a similar mechanism—when the ambient air around them is dry, the scales that hold the seeds spring open. If it's damp, however, they clamp shut. Mechanisms like these usually depend on two different layers of tissue that tug in opposite directions as they absorb or release moisture. Humidity exposure, in essence, is translated into mechanical motion, not by separate components, but by sensor and actuator combined in a single chemically active sheet. No brain decides when pine scales should open and close; no nervous system is required to signal the change.

The artificial leaves operate on the same principle—when water vapor increases near the chemically sensitive polymer on the surface, the leaves contract, bending toward or away from the source in response. Drying them out, on the other hand, causes the reverse motion—they can bend in both directions. This movement requires no energy expenditure from the Plantoid, which is a big deal in robotics. HyQ was a powerful beast, but it was also power hungry—to run its hydraulic systems, to canter and trot, required an external source of electricity or a hefty load of batteries. As Mazzolai put it: "Most of our robots just work for one hour and poof! They're done." A passive actuator, however, operates without such energy requirements, even if its repertoire is limited.

Del Dottore motioned for me to try it. I dipped my finger in the vial and raised it toward the surface of the leaf, letting it hover there. The "leaf" began to curl down, as if it was wilting in real time. It brought to mind a memory of a "sensitive plant" that my childhood friend's mother kept in a pot on a windowsill. When we poked this plant's leaves, which were arrayed in a series of small feathery lobes on either side of the stem, they would pinch themselves together like a row of praying hands, begging us, in essence, not to do it again. It fascinated us so much, in fact, that the mom eventually intervened—stop touching the plant before you kill the thing!

What we liked, I think, was the strangeness—a plant responding to us, defying all our expectations. It was a glimmer, I think, of what's so strangely appealing about the idea of a robotic plant, some seemingly alien potential unlocked in the familiar figures of leaf and stem.

~

When my family and I got back home from the United Kingdom, the garden was no longer the same. It was August, the doldrums for growing plants in South Carolina—there were no bamboo shoots emerging from the parched lawn. But what was different lay deeper, in how I viewed the botanic presence that surrounds our house. The garden was no longer a still life. Del Dottore had mentioned the movement of vines as a new research area for the team, and I immediately thought of "the vine that ate the South." Actually, there are at least half a dozen vines that are busy eating the South, and they aren't kudzu. Greenbriar, for instance. From a giant white tuber in the ground, thorny vines rocket straight up into the canopy each spring, the stem stiff enough to support a vertical trajectory. Then there's the loops of wisteria. Then jessamine. And English ivy. Clematis. Virginia creeper. Muscadine grapes. Trumpet vine, bittersweet, two species of passionflower including *Passiflora caerulea,* whose tendrils are the model for new Plantoids, and honeysuckle. Each has its own clambering strategy, and also its own adhesive strategy. Ivy, for example, has something that looks like suction cupped paws that cement themselves to the bricks of our house. Jessamine winds and drapes itself over branches but doesn't adhere. Grapes have soft green tendrils that curl around branches and then harden into coiled wire as they brown.

In our neighborhood, you'll see vines like this that got away at some point, the wisteria wrapped around the base of a loblolly pine like a python, then sprawling way up in the canopy. You'll also see the remedy—cutting the vines at the base leaves a thick wrapping of dead fiber, like the husk of a giant coconut, the layers dwarfing the actual host trunk.

After a month away, our yard was headed that way; in fact, the bamboo stand looked like an arbor. Talk about entanglement! If the underground realm is the rhizosphere, our subtropical backyard has got to be the vinosphere.

They were all moving, these vines, in their peculiar, nonhuman way. Even as I stood still in the shade one afternoon and stared at them. Living examples of the subjective experience of time.

2

INTERPRETIVE STATION

Bats and biomimetic sensors

When I was in seventh grade, I stayed home alone one day with the flu. It was late fall, and the leaves had dropped from the hardwoods and were swirling around the yard in gusts of unseasonably warm air. The weather was pleasant enough that I was sitting with the sliding screen door open to the deck, reading.

Every time I heard the breeze rise, I would look up from the page and watch, in my feverish stupor, as a flock of leaves revolved around the corners of the deck, tumbling and twirling as if they were alive, until the wind died down and they drifted into an inanimate pile of foliage again.

This kept happening, this contrapuntal pattern of a turned page and a scuffle of leaves. I watched one flock tumble all the way up the driveway, past the basketball hoop, toward the house. They blew through the deck railing, crossed the deck, and hit the screen, pressed to the mesh like trembling fish. My eye traveled down to a hole poked in the lower corner, where a dog had once pawed through. I wondered, listlessly, whether a leaf would fit through there, and if so, would I be obligated, having seen it happen, to stand up and throw it out again? I hoped not. I didn't feel like moving.

I watched, mesmerized, as one crinkled brown leaf hit that hole...and twisted through it, rolling onto the floor. Okay, I thought. That answers the first question.

Then, it seemed to elevate. Was it—flapping? It was indeed. Up to the ceiling it rose, circling the room once, then twice. Over my stunned face, my open mouth and open book, it revolved in that distinctive wobble and swoop you normally see silhouetted against a dusky sky. Then it broke off and continued its tremulous flight down the hallway to the bedrooms, where it disappeared.

Cautiously, I closed the book, as if it might contain something even more portentously ironic, something that shouldn't be disturbed.

The book was Bram Stoker's *Dracula*, of course.

I called my dad at work.

There's a bat in the house, I said. It got in through the hole in the screen door!

He asked if I'd taken my temperature recently. Maybe I should take some more aspirin.

I'm telling you, I said. I saw a bat. It came in the house and it's still inside.

Eventually, my dad came home to check. I showed him the hole. I mimed the whole thing, my hand fluttering overhead, pointing to the bedrooms.

It went that way, I said.

He pushed clothes hangers around in the closets. He inspected the corners of the ceiling and peered behind the doors and in the cracks behind the dressers. He actually did a pretty thorough survey before he looked at me indulgently. Was I sure I wasn't napping? Maybe I should try reading a different book. He was in full father-knows-best mode as he got back in his car.

Maybe I *was* napping, I thought. Maybe I'd been in one of those twilight states where you reach the end of the page and realize you haven't registered a single word. As the afternoon passed with no further appearances, that seemed more and more likely. But it seemed so *real!*

By evening, my fever had risen again. Darkness fell outside. I was sprawled on a makeshift bed on the floor of the living room, too delirious to focus on whatever was playing on the television. Not reading *Dracula* or anything else. My dad and his girlfriend Nancy were in the kitchen making dinner. I heard them laughing about my story in there, and I was too sick to protest.

They came out with plates loaded with tofu, rice, and lentils. There was something they wanted to watch; *Cheers* probably, or *WKRP in Cincinnati.* They plopped themselves down on the couch, but instead of monotonous chewing, I heard a startled titter, then a curse and an abrupt calamity of silverware.

Looking up, I saw my dad bent over with his head squeezed between his shoulders, trying not to spill all his food on the bile-hued shag carpet as he waddled toward the kitchen. I saw Nancy crouching on all fours, scrambling behind him. And there was my imaginary friend, dipping and bobbing in circuits around the room.

I felt oddly detached from the excitement—if a winged elephant had emerged from the bedroom and started careening around the popcorn ceiling, I don't think I would have moved. I couldn't express myself with anything more than a pointed finger and a cluck that passed for exultation. The adults weren't laughing now.

But what was it doing in there? What was it thinking? Was it sick too? It made more sense as literary hallucination, although in truth, there really weren't that many bats in the book.

My dad came back with a broom. Nancy opened the front door wide. I watched as they ran around and around in this strange carousel, swatting and shrieking until the bat finally escaped into the night.

~

What is it like to be a bat?

That's the title of an essay by the philosopher Thomas Nagel which, although published in the early 1970s, has become something of a contemporary touchstone in the study of all things nonhuman. There's Nagel's bat fluttering across the pages of J.M. Coetzee's polemical novel *Elizabeth Costello*, and there it is again in Jenny Diski's musings on "the animal question," and then again in *Being a Beast,* in which the naturalist Charles Foster swats at it with a touch of impatience before embarking on an attempt to become a badger. It flies again in computer scientist turned philosopher Ian Bogost's contemplation of the nonliving world, and implicitly, it's flapping around in the backstory of Derrida's portrayal of his ogling cat, "an existence that refuses to be conceptualized," and in the performance artist Eduardo Navarro's impersonation of a Galapagos tortoise named Lonesome George, and even in Thomas Thwaites's capering about in a goat-inspired exoskeleton in *Goatman*. "What is it like to be a fungus?" asks the biologist Merlin Sheldrake in a recent book about the hidden mycelial universe. It seems everybody these days wants to know what it's like to be nonhuman.

But how?

Nagel answers his own question pessimistically: we'll never know. "Even without the benefit of philosophical reflection," he writes, "anyone who has spent some time in an enclosed space with an excited bat knows what it is to encounter a fundamentally alien form of life." Bats, he acknowledges, appear in his essay as a rhetorical convenience, because their distinctive "sensory apparatus," their use of echolocation to orient themselves in the world, offers such a sharp contrast with our own tools of navigation. Beyond that, however, he's not really interested in exploring their worldview. "It will not help to try to imagine that one has webbing on one's arms," he writes,

> which enables one to fly around at dusk and dawn catching insects in one's mouth; that one has very poor vision, and perceives the surrounding world by a system of reflected high-frequency sound signals; and that one spends the day hanging upside down by one's feet in an attic. In so far as I can imagine this (which is not very far), it tells me only what it would be like for me to behave as a bat behaves. But that is not the question. I want to know what it is like for a bat to be a bat.

Sorry, aspiring nonhumans; you can take off the animal suit.

Nagel's real interest lies in the limits of "reductionist" science to explain consciousness, bat or otherwise. If bat subjectivity can never be traced to its foundation in particle physics, then it suggests that science itself cannot provide a universal explanation for the way the world operates. There are phenomena out there that are impervious to our conventional methods of inquiry, our attempts to describe the world according to a set of foundational laws. As Bogost puts it in *Alien Phenomenology*: "To comprehend the effects of the high frequency vibrations voiced and heard by bats simply has nothing to do with understanding what it's like to be a bat."

Having spent some time in an enclosed space with an agitated (and probably not *excited*) bat, I find myself agreeing with Nagel to a degree: in our efforts to experience battiness, we wind up learning about ourselves, our own sensory limits and subjective obsessions, our own nature as a species. The fact that I can't know all about alien creatures is what makes them so endlessly fascinating—it makes me pay more attention to the particulars of their existence, not less. But I also think there's something important in the phrase "what it is like." We can't know what it *is* to be a bat bobbing around the ceiling. But the mimeticity of what it's *like* to be a bat? That I'm curious to explore. Bogost arrives at a similar approach: "We never understand the alien experience," he argues, "we only ever reach for it metaphorically." The question, however, is what kind of metaphor, or metaphors, reveal the most about these other beings, and ourselves. In my view, the practice of nature writing involves the creation of metaphors that are multiple and interdisciplinary. It is a broad kind of metaphorism, one that ironically includes scientific inquiry within its repertoire even as it tends to chafe against the limits of its ideology and methods. Science, in other words, can help forge new metaphors.

In particular, biomimetics is predicated on the idea that we can know what it's *like* to be a bat, not literally, but figuratively, by accumulating the detailed natural history of bat life, then coming up with a model that represents its defining features, and finally representing that model in physical terms. We speculate about the bat's existence; we imagine, based on the details we've gathered and the models we've made. It's an inherently interdisciplinary approach, one that doesn't have to be limited to scientific inquiry. Indeed, my concern is that in dividing the appropriate turf of the sciences from the humanities, we reinforce a divide between nature and culture and leave a lot of terrain unexplored. As biomimetic practitioners, or nature writers, we might proceed along the lines of Coetzee's fictional novelist, Elizabeth Costello, who responds to Nagel's question with the triumph of humanist imagination: "If I can think my way into the existence of a being who has never existed, then I can think my way into the existence of a bat or a chimpanzee or an oyster, any being with whom I share the substrate of life."

Instead of just thinking our way in the abstract, however, I'd say we also need to walk the walk—the "substrate of life" is yet another landscape metaphor, inviting us to get out there and explore the real world, our best model for what's going on with other species. The more we get out there and explore all the layers of that substrate, the more precise our metaphors become. That bedrock motif also invites us to extrapolate beyond the living, as object-oriented ontology does in asking what it's like to be a nonliving object, like a computer. What's out there on this trail is alive and not alive. To craft a biomimetic creature is to build a model of these relationships between all these different entities—the soil is implicit in the way plant roots have evolved to burrow, for example. The biotic doesn't exist in the abstract; it's embedded in the broader material world.

More humbly than Coetzee's fictional character, we might acknowledge that neither the literary nor the mathematical model is the real deal, the genuine chiropteran. The webbing and the algorithm are different forms of modeling, tools to sharpen the figurative imagination. We tend to think of these approaches as exclusive and proprietary—it has to be science or poetry, and only one can be the credible way to represent the existence of another species; the other is just superstition, or "magical thinking." We like to think that only the living matter, or in rejoinder, that actually the nonliving are ultimately what count most. But the biomimetic approach suggests that the substrate is actually common ground: we speculate about what it's like to be a bat by making multiple versions of imitation bats, poetic, technological, and interdisciplinary, and then hopefully weaving them together.

We already know enough about the substrate of bat existence to know that echolocation renders the same world differently for each species, and indeed, the same is true for all the myriad ways nonhuman species sense the world. For us it's an invisible geography, yet it's thronged with other creatures, a substrate of life that is, paradoxically, right there and yet beyond us, opening the space for metaphor. For this stretch of our nature walk, imagine the world around us as we don't know it—chirps and trills emanating and reverberating between the "noseleaves" and ears of horseshoe bats, sketching landscape portraits with brushstrokes of sound. Imagine too the wafting particles of scent that a truffle exudes from underground, powerful enough to lure rooting hogs, mingling with the pheromones of moths and the carrion smell of pawpaw flowers, a beacon for tiny flies. Then there's the flare of infrared light radiating from petal constellations to signal bees, and the tremor of an elephant stomp radiating through the ground, and the magnetic orientation of the planet, captured in the optical organs of songbirds migrating at high altitude through the night, feeling their way across the continent.

It's easy to look past the "sensory apparatus," to view the eyes as the window to the soul, just a transparent way station on the path to what really matters inside the brain. The brain as the treasure chest of ultimate answers—that's an anthropocentric construct, one that biomimetics challenges by posing alternative visions of where meaning lies. Consider Brooks's contention that sensing is actually part of the essence of being—in bats, significant interpretation of acoustic soundwaves happens in the cochlea, before it reaches the brain. Plants have no central brain, yet they have what amounts to vegetal intelligence, their processing and reacting abilities intertwined with their sensory abilities. The sensory apparatus is more than just physical equipment; it's a process of engagement. Echolocation involves both the bat and the landscape—what you're modeling is the process of the individual responding with its senses to the rest of the ecological community.

On this stretch of the trail, then, we'll give the senses their due. We'll get to know what it's *like* to be an echolocating creature by imitating the way a bat experiences the world. We'll go there even if Nagel himself wasn't so keen on the idea.

~

In my various encounters with literary bats, *Dracula* included, I've always felt a twinge of disappointment: they all seem to be sketches of a generic bat, as if it's possible to lump all these winged mammals together and file them under a few key terms: echolocation and phobia for starters, and most prominently, bloodthirst. No avid birder would be satisfied with "a bird," but there seems to be a common assumption that a bat is a bat, and that's that. Not so for me: what's always irked me about my teenage encounter with that little brown job was that I never knew, and would never know, what kind of bat it actually was. Was it, perhaps, the once common little brown bat, *Myotis lucifugus*, known for its habit of roosting in human habitations? Or a big brown bat, *Eptesicus fuscus*, also a denizen of attics and wall crevices? Or maybe it was a *Lasiurus borealis*, an eastern red bat, which sometimes hibernates on the ground beneath deciduous trees, blending in with the fallen leaves. It was long gone before I was well enough to even think about making any kind of identification, not that I had a field guide to bats available to me at the time. It would remain forever just "a bat."

In reality, of course, there isn't just one generic bat species flying around—there are multitudes. There are Old World and New World bats, "megabats" with three-foot wingspans and "microbats" the size of an unshelled peanut. Twenty percent of the mammal species on the planet are bats, and they can be found just about everywhere but the poles. Thirteen hundred chiropteran species isn't so impressive, perhaps, when compared to the 2,000 species scurrying around in the order *Rodentia,* or the 6,000 species of passerine birds, or the 360,000 known species of beetle, but I'd argue that by and large the physical diversity among bats is as perceptible to us any cardinal or warbler, if we actually get the chance to look closely, face-to-face.

We now have galleries of bat faces to ponder online, which I never had as a kid, and those visages make it vividly clear that a bat is not just a bat. The little brown bat has a flattened face, for example, like a tiny pug, while the big brown's face is more of a blunt black wedge, tipped with elfin ears. But those are relatively subtle differences; worldwide, the faces of bats run the gamut from reasonably cute to hideous. "Flying foxes," for example, have big soulful eyes that seem to have evolved more for the sake of cartoon animation than finding fruit. Like an archive for horror film inspiration, many others seem to have rejected every detail of human comeliness. The eyes are often an afterthought, but the ears are crenellated and warped, stretched to a point, or mashed grotesquely around the chin, while the mouths are surrounded by fleshy wet growths, all phantasmagorical folds and whorls and grooves. Once you get over that, there's often a last flourish of demonic spikiness somewhere on the snout. If you're looking to dispute the idea that there's "a bat" out there available for rhetorical deployment, take a look at a gallery of these faces.

Not only are there well over a thousand variations of what it's like to be a bat, but that diversity must mean something for the way these different species exist in the world—their facial features reflect different natural selection pressures, different environments, different behaviors. Why, for example, are some species, like the greater spear-nosed bat, *Phyllostomus hastatus*, adorned with a rhino-like spike called a lancet on the tip of the nose, while others, like Rüppell's Horseshoe Bat, *Rhinolophus fumigatus*, appear to have something like a multi-necked violin adorning their nostrils, which biologists refer to as a "noseleaf?" And why such vast differences in the size and shape of the ear? One species, Wagner's mastiff bat, *Eumops glaucinus*, sports what looks like a wrestler's crushed lobes enveloping the sides of its head, while *Corynorhinus rafinesquii*, Rafinesque's big-eared bat, appears to fly around with a split avocado peel thrust vertically above its brow. Some have inverted, dish-like faces; others have protruding muzzles. What purpose does all this gargoyle embellishment serve?

One answer lies with echolocation, or biosonar, the distinctive apparatus that makes being a bat so different from being a human. Biosonar involves a variety of physical features, but it also can be thought of as a process with three distinct characteristics. Detection is first—sensing something's out there. Then comes localizing it in space, determining distances and angles. And then there's identifying, or classifying, the distinguishing features of shape and texture and movement pattern. To understand how it works, we have to start with an acoustic sense of the landscape, and how sound moves through it in an oscillating pattern of crests and troughs. If you've ever floated around on a surfboard waiting for a set of waves to arrive, you'll probably have a sense of how this works. What we hear is partly a reflection of the size of these waves—amplitude is the term for measuring how large the wave is, how much energy it contains. Bigger is louder. Then there's wavelength, which refers to the breadth of the troughs between the waves. The term *frequency* adds time to the equation, since it describes how many waves cross a specified point in a specified time. The higher the frequency, the closer the waves are together, and the shorter their wavelength. A high-pitched sound is produced at high frequencies; the lower the frequency, the deeper the pitch.

The human hearing apparatus is limited to a certain range—whether we register a sound depends on the amplitude, or how loud it is, and on the frequency, since there are noises that are too high and too low in pitch for us to perceive. Sometimes our spectrum overlaps with other species, like cicadas trilling from the treetops or the eerie flute notes of a wood thrush emanating from the forest floor, and sometimes, as in the case of bats, these ultrasonic clicks and shrieks may be well beyond the high end of our range.

Echolocation is a call and response procedure—it involves sending out sonic waves and then comparing the original pattern to differences in the waves that come back. The bat sends out pulses of sound in a "beam," either from its nose or its mouth, and then monitors the airwaves for reverberations as the waves strike an insect or the wall of a house and bounce off, returning to the sender but altered in shape. It's similar to what we do with a flashlight: we send out a beam of light and see the light that comes back at us from whatever it strikes in its path. Far away objects glimmer dimly as the cone of light widens and dissipates; close objects are awash with glare. We do our best under these conditions; for us, the flashlight is a contrivance we use to keep from stumbling around in the dark. For bats, however, biosonar offers the chance to seize the advantage over more visual competitors. In the dark above our heads, a bat detects its prey, pinpoints it in space, and continues to discriminate the insect from the static of whatever else is floating around out there even as it swoops in for the intercept. There is, after all, a lot of noise out there on a summer night. Think of the katydid chorus rising and falling, add in the chirrup of the treefrogs and the cockroach and cicada trills and all the ultrasonic chirps and clicks we can't even hear—then imagine all that ambient noise was light instead, coming at you like headlights from all directions.

"Hawking" mosquitoes in an empty sky is a different proposition than "gleaning" a katydid from a leafy bough. It would be one thing if the bat stayed immobile, like a lighthouse, and its prey stayed fixed in one spot as well. But because bats are often on the move and their prey is too, they have to adjust to what's known as the Doppler effect: by moving closer to the target, they are compressing the wavelengths in front of them and changing the shape of the echoes they receive. They also have to account for the movement of their escaping prey, to the sound signature of each echoing wingbeat.

Some bats use what's called a "constant frequency" strategy to make these adjustments—they constantly adjust the frequency of their calls to compensate for the impact of movement in the echoes they receive, and they often do so according to a species-specific harmonic pattern. They might slow down the rate of the sounds they emit as they approach a target, for example, spreading out the frequency of the waves in advance to account for the compression that happens as the target gets closer and closer. That constant focus on a limited frequency is like using a flashlight with a narrow beam—it zeroes in on a small area while eliminating the "echo clutter" of the surrounding background noise. "Frequency modulated" calls, on the other hand, sweep through a range of wavelengths, widening and narrowing the beam the bat is following, much like we might rotate the head of a flashlight to scan a wider periphery with less intense light. Constant frequency calls reduce the breadth of information; frequency modulated calls pick up a wider range. One offers focus, the other context. Many horseshoe bats utilize both strategies—they lock in on a target in motion using constant frequency calls, but they also emit a frequency modulated chirrup at the end of a call, widening their sonic beam to give them a sense of potential obstacles as they delve into the undergrowth, making sure their prey can't suddenly swerve to the side and disappear from the beam.

Swerving and diving—that's just one maneuver in a potential prey item's repertoire. While bats have been refining their locating strategies via evolution, their quarry hasn't just been sitting around, oblivious to everything but moonlight and pheromones—they know they're being stalked. Comparative studies of related moths have shown, for example, that a moth species on the Hawaiian island of Kauai, which has only a single species of bat to worry about, has adapted its hearing to tune in to the frequency used by just that one adversary. Meanwhile, on an island off the coast of Panama, a related moth species has had to cope with predation by several different bat species, each with a different signature sound. These moths have evolved to stay tuned to several frequencies instead of just one. Insects that hear they're being targeted are also adept at measuring how close the danger lies. A moth may turn and flee if it judges the threat is far enough away, or zigzag then drop into a freefall if the bat is getting close.

Bats have evolved beyond just modulating the width of their beam in response: they might produce a pulse of sound then shut down their hearing for an interval to screen out white noise, like the burst of clicks some moths produce in an attempt to create "acoustic camouflage," or the ultrasonic din that might be coming from other bats competing for the same food. Reading about these different scenarios, the night sky starts to seem more like a traffic jam or a battlefield than the quiet abode of twinkling stars.

While it might seem that echolocation offers a bat centric principle of sorts, even here it's difficult to generalize. There's been quite a rigorous debate, in fact, over whether echolocation evolved just once and was so successful that it radiated outward to create greater and greater diversity around the globe, or if it in fact happened more than once in different locations, with convergent evolution pushing these ancestral creatures toward what looks like a shared set of characteristics. The differences in bat facial features attest to just how finely tuned each species is to a specific ecological context: bats look different because they lead very different lives. Species that catch their victuals on the wing, at high altitude, need to be able to pick up low-intensity sounds emanating across long distances—they need big ears to hear what's going on. Gleaners often wait until they get very close to the surface before they send out calls, since what they need is a precise map of what's right in front of them. Some species broadcast their calls using their mouth; others emit a sonic beam from their nose, then modify it further using all the crinkles and lobes and grooves of their noseleaf, a procedure known as "beamforming." Not all these differences can be traced to the capturing of prey: there are bats that feed almost exclusively on nectar and pollen yet still sport a spiky lancet on the tip of their nose, like the Mexican long nosed bat. What looks frightful to us is actually a finely tuned instrument, the purpose of which we may not have figured out yet, there being so many different permutations.

As I delved into bat biology and technology, I soon discovered someone whose research seemed to be exploring similar terrain, albeit from a different disciplinary perspective: Rolf Müller, a professor of biomechanical engineering who runs the Center for Bioinspired Science and Technology at Virginia Tech. Müller's research focuses on bats, but more broadly, he's exploring technology as a reflection of biodiversity, which is akin to what we're doing as we amble along this biomimetic trail, looking at biodiversity through the lens of technology.

When I called him up, Müller offered a pitch for what makes bats so distinctive and valuable as a model: their diverse nose and facial features, he said, are part of a dynamic sonar system that operates far more efficiently than the conventional sonar systems we've devised. A submarine, he told me, typically deploys a vast array of microphones and enormous dish receivers to concentrate signals into a narrow beam that scans the surroundings. Bats have just two receivers, their ears, and one emitter, either the mouth or nose. But by adjusting the structure of the ear or the nose, they can change the width of the beam and its pattern both when they release the sound and when they receive it, strategies that allow them to do more with far less. He'd made a similar point in an article I'd read, contrasting the perception of big brown bats with a conventional imitation. In an experiment, trained big brown bats were able to maneuver between hanging brass rods that were only 1.5 degrees apart. If a conventional sonar array with similar ability were to be built, he and his coauthors had written, "the result would be 80 times the size of the animal and contain two orders of magnitude more elements."

The goal of Müller's biosonar research, one might say, is to figure out how bats do it better by identifying all the nuanced expressions of sensory biodiversity in bats—all the distinctive deformations and flaps of the ear, as well as all the fleshy, contoured noseleaves with their distinctive notches, folds, furrows and lobes—and translating them into models for useful technological designs. Not *a* single bat-like device, but many. This work has taken him in some surprising directions for a professor of mechanical engineering: one of his recent projects created 3D scans of the fluid preserved specimens that make up the Smithsonian Museum of Natural History's voluminous bat holdings, some of which may have been collected in the field decades ago. He also does a lot of globetrotting to gather his own anatomical data, bushwhacking and spelunking in remote corners of the tropics where the highest concentrations of *Rhinolophidae,* the facially flamboyant family known as horseshoe bats, can be found.

Over the phone, Müller told me the story of how he'd developed his interest in bat ear shapes. He'd studied animal physiology at the University of Tübingen in Germany, and he'd set up a cooperative research station with Shandong University in China to conduct research on horseshoe bats, including the functionality of noseleaves. His field research usually involved finding specimens in the wild and taking video footage of their faces, which could then be translated into data sets of noseleaf geometry. On one of his tropical cave visits, however, he noticed something else was happening when he reviewed the video footage—with some species, the images were always blurry around the pinnae, or ears. It was an "a ha moment"—the ears were moving, flapping, and pinching and rotating at 1/60th of a second, three times as fast as the blink of a human eye. Something was happening, in other words, not just with the way the soundwaves emanate from the mouth and nose, but with the way they were received through the changing shapes of the ears. Müller consulted a German study of bat anatomy from the 1960s—bats, it turned out, have 20 ear muscles. We have four.

"Nobody cared about these things," Müller told me. "I kind of fell into the question."

Back in the lab, Müller began converting the video footage into data sets of noseleaf and pinnae geometry, a catalogue of possibilities for "beamforming baffle shapes" that those in his lab refer to as the "nose zoo" and "ear zoo." Biodiversity became technological diversity. There are now over a hundred digitized models in each of these collections, and Müller's lab has been developing a series of biosonar models based on them, even strolling through the woodsy parts of campus with their bat-like device, holding it aloft while it chatters ultrasonically through its noseleaf, like a horseshoe bat. This was a creature I wanted to see, even if I couldn't visualize, or even hear, much of what it was actually doing.

~

I arrived in the quaint college town of Blacksburg, Virginia on a chilly afternoon in mid-February—snow flurries and black ice were in the forecast, and an expectant, misty hush had draped itself over the hills surrounding campus. Like some ironic echo from seventh grade, I was once again suffering from the flu, a mild case this time but one that nevertheless meant tissue collections in every pocket and a general malaise I hoped might somehow be inspirational—didn't Coleridge emerge from a fever dream with the stanzas of "Kubla Khan" flowing from his pen?

A full day of activities lay ahead. The next morning, I grabbed a cup of coffee and some ibuprofen and dashed to an early class Müller was teaching on biomimetic principles in engineering; the day's topic was optical sensors, based on the way that insects, sharks and other creatures have approached the concept of seeing through very different eyes.

Müller, who is tall, with carefully parted brown hair and eyes set wide beneath a prominent brow, had a gentle way of coaxing answers out of the still drowsy students. After posing a question in his faint German accent about the trade-off between time and spatial resolution in pinhole cameras, he waited with a wry indulgent smile, knowing, with a veteran teacher's confidence, that someone would fill the pause.

Afterwards, I gave a talk myself, to a room full of engineers and computer scientists gathered for their weekly biomimetics colloquium, which for an English professor was an unusual undertaking. I tried to recall some of Müller's poise. Then we traveled to his lab on the edge of campus, on the other side of a pond thronged with waterfowl, where I sat down with two members of the team, postdoctoral researcher Omar Khyam and grad student Liujun Zhang, to talk about modeling the way bats process the information they generate and receive.

The bat-like creature sat on the table between us, its sensory priorities immediately clear. What was striking about Semini's quadruped robot HyQ was it appeared to be a dog with no head; it had no functional need for one. Plantoid had no bean either; it's sensory and intellectual capabilities were distributed from branch tip to root tip. Echolocation in bats, on the other hand, is all about the head. The "acoustic fovea," the technical term for the special region of the bat's inner ear where sound is processed, and all the essential componentry is located either on the head or inside it. The biomimetic version followed suit: it was an enlarged version of a greater horseshoe bat head, clamped to a plastic box full of small electric motors, circuitry, and wiring. Although it had no cameras to serve as eyes, it appeared to have a face, albeit one devoted to sensory functionality rather than emotional expression.

Two ears, enormous by bat standards, were attached in horseshoe bat position at the brow, above a noseleaf that looked like the inside of a cracked walnut, surrounding the aperture that emitted ultrasonic pulses and tones. White and fleshy, the 3D-printed features selected from the nose and ear zoos looked convincingly biotic; while everything else looked rectangular and linear and mechanical, they looked as if they'd been grafted from a live donor. Behind the microphone in each ear was a plywood lever, tipped in grey foam, that was connected to a small motor. When activated, it was meant to quiver like a muscle group, striking the back of the soft ears and "deforming" their shape in the manner of a horseshoe bat capturing echolocation signals coming back. The team had also hidden miniature pneumatic actuators—tubes delivering pressurized air—behind the ears as well in some versions, to enable the variety of rotations and shape changes that an actual bat might deploy in each ear. There were rods and motors behind the noseleaf too, pressing it down to increase its "concavity" while shaping the pulses coming from the loudspeaker within.

Khyam said their work was part of a much larger study, funded by the Office of Naval Research, on the application of bat sonar to drone navigation. The overall goal was to devise a model for the way these tiny mammals are able to navigate so effectively in the midst of dense vegetation and apply it to a control system for drones operating in the cluttered space just above street level. Bats offer an existence proof: we know the solution exists because we know bats can do it, and that's with just two grams of brain matter, slightly less than a penny. In fact, we know that for some species, environmental clutter is what they do best: a hunting horseshoe bat might look for certain species of trees that it knows are favored by some moth species; it might use the beam patterns of leaves along the edge of the understory to identify a landmark it can follow, or a habitat where edibles congregate. One of my favorite symbiotic success stories involves *Kerivoula hardwickii,* Hardwicke's woolly bat, and a tropical species of pitcher plant that is normally carnivorous but has evolved to subsist on bat dung. How does it get enough of this fertilizer? By providing a safe space for these bats to roost inside the pitcher, which it advertises by reflecting a specific pattern of echoes from its foliage, like a hotel billboard.

This talk of bat heads mounted on drones brought to mind Claudio Semini's early work with HyQ, how when they added cameras to allow the robot to "see," it was with an eye to *what* the sensors could gather, not *how* they gathered it. Here was a similar reckoning with real world conditions, but now the gathering process was the key. Although Müller is also researching the dynamic flight possibilities of bat wings, this part of his team is teasing out the mechanics of what it's like to sense the world as a bat.

Khyam described the "spike model" they'd been developing, based on the assumption that bats use a pathway for translating echoes into neural "spikes" that is similar to human hearing. Tiny hairs inside the ear vibrate with the incoming sound pattern, triggering a flow of calcium ions that fires nerves at the base of the hairs—the resulting "spike trains" travel along neural networks into the brain. The bats are converting sonic data into representative models, written in biochemical and electrical codes. The research team's models operate in similar fashion, except the cochlea is simulated by a silicon chip and the spikes are simulated in code, meant to traverse artificial neural networks.

"Right now, we don't have a better model for complex environments," Khyam said, offering a cautious appraisal of their progress, as we looked at sonograms that resembled potted plants tipped sideways, conical at the start, then jagged and spikey before tapering off in a single stem.

"But we've found the whole process so far really mimics the bats."

Designing an artificial pathway for this translation of sonic stimuli is one step; the other is to figure out what kind of data is actually traveling down that pathway as the bat interacts with the landscape. The basic assumption is that bats are not operating randomly in these environments; they're not just crashing through the undergrowth. There must be patterns they're recognizing, based on differences in the signals between one area and another. But what would be an accurate model for the "beam patterns" we know bats must be detecting when they operate among the leaves? What does a leaf sound like to a bat? What about hundreds of leaves in a forest composed of pine, magnolia, and maple? Müller and his research team have been studying this problem for over a decade, using different simulations to model leaves in space—flat and smooth circular discs, for example, or foliage models that take into account the average radius of the leaves, their orientation, and their density. Recent models then add a layer of branching representation according to what's known as an L-system model, named after the Hungarian botanist Aristid Lindenmayer, who developed an algorithm for generating representations of branching patterns in plants. The results look like elegantly stylized pencil sketches.

To test the accuracy of these representations, they applied echolocation pulses from actual bats to the virtual leaves. The pulses were drawn from the greater horseshoe bat, *Rhinolophus ferrumequinum,* whose noseleaf isn't the most florid of possibilities. In fact, as these things go, it's fairly demure, just two fiddle-like lobes encircling the round disc that contains its nostrils, then a pinched arrowhead perched where the nose would normally be, beneath a series of bumps and crosshatching ribs heading up the flattened snout to the pointy apex of the lancet. These simulations generated a series of simplified foliage beam patterns, which could then be compared to how the biosonar head performed out in the field with actual leaves. The first of these alfresco experiments was stationary—the biosonar was mounted on a tripod at various angles to survey a particular tree. The next step was to put the echolocation device in motion, like a greater horseshoe bat on the wing.

Out in the woods, this mobile creature generated an extensive "foliage echo data set," recording the horseshoe bat-inspired call patterns as they bounced off trailside vegetation, then pairing that acoustic data with visual reference points gathered by two Go Pro cameras. The ultimate goal was to create a data set big enough to employ machine learning to train the control system to recognize various beam patterns so it could identify and respond to similar conditions as it maneuvered through the vibrant chaos of the great outdoors. To gather this big data set, Khyam and Zhang actually spent hours traipsing through four different forest types on campus, occasionally turning the batbot off when the motors got hot enough to burn their fingers. Although I could imagine the intriguing juxtaposition of greenery and whirring echolocatory robotics, we were not going to replicate that walk in a leafless forest peppered with freezing rain. Given my physical state, I wasn't altogether sad to be exploring the indoor habitat instead.

While we were talking, a bit of engineering commotion was happening in the lab across the hall. Each year, students in Müller's courses try to improve the design of the batbot, addressing the shortcomings of the prior version with new and hopefully better ideas. The design team was meeting later in the day to review their progress, and a lot of last-minute tinkering was going on in there.

Joseph Sutlive, a graduate student who jokingly referred to himself as "the sonar head guy," led me inside. Several students were congregating around what looked like the kind of equipment you'd see in a sound studio, watching a pattern of wavelengths on a laptop. Tweaking something. The lab benches were covered with a strange assortment of materials that suggested engineering work in the digital age—wire snippers and snippets of wire, circuitry boards, but also the detritus of biomimetic bats—ears were lying around, some hard, some rubbery, and a handful of nose leaves were scattered on the lab benches too. In one corner was the 3D printer that produced these features, and in another was a curtain of plastic vines hanging down to the floor from a suspended wire—artificial habitat for artificial bats.

Sutlive pulled open a drawer—inside were more bits and pieces of imitation bats, some with embedded emitters and wires still attached. One of his tasks had been to translate the images of bat pinnae into the digital models in the ear zoo using an animation software that allows for the creation of 3D models based on measurements captured from high resolution photos. That meant scanning the image and establishing points on the surface, then connecting the points to create a simplified mesh. This idealized version eventually made its way to the printer.

"We've got the noseleaf down pretty well," he said, "but the ear is still a work in progress."

He rummaged around and handed me one—it was molded from hard black plastic, with a pointy tip and a cube at the base where the mold was held in place during printing—it looked like you could probably scoop ice cream with it. A spine, corresponding to the cartilaginous helix that rounds the outside of our ears, ran vertically to a curved nib along the upper edge, but otherwise there were none of the details of horizontal ribbing you might find in an actual horseshoe bat ear. Sutlive said they'd developed a series of static versions like this, then switched to more flexible materials, based on the need to mimic the way a bat can make each ear rotate, pinch, and deform independently.

"It's more than a simple bend," Sutlive observed drily of the challenges posed by soft robotics.

Around the corner from the curtain of plastic vines sat a Halloween version of the batbot. It was similar to the one Khyam had just shown me, but all of the features were rendered in a soft black polymer that looked pretty worse-for-wear, like Darth Vader's half-melted helmet that reappears long after his death. Unlike the hard plastic ear I was still holding, the cups of these pinnae were ribbed horizontally, providing a wavy surface that creased the ear to allow it to fold under pressure. There were four of these rounded bulges on the imitation, and although I recalled more of them on the actual bat, what caught my eye was the way they curved like fish bones right up to the tip of the ear.

A bundle of multi-colored wires ran from the innards of this creature across the bench to a fairly primitive looking device for measuring the ultrasonic wavelengths coming from the nose. Sutlive pulled one of these wires and connected it to a laptop, then entered some commands to power up the "reptilian brain" that worked with the commercially available microcomputer they'd installed to power the control system.

"It's on," he said.

I waited. There was an audible click coming from the speakers behind the noseleaf, like someone flicking a lighter, or the sound the needle makes at the end of a vinyl LP. It wasn't the actual pulses, however. Sutlive said those are usually filtered at the receiving end to create a "hard window" around a certain frequency range and prevent high frequency noise from scrambling the signal. Many bats species do this too—they have a biological "filter bank," the equivalent of the equipment you'd find in a sound studio, to "decompose" the incoming sound waves and exclude disruptive noise from the echoes.

"Is it actually on?" I inquired. It still looked like a mask waiting to be worn.

"Oh, it's on," he said. The ears were just in static mode, picking up signals but not moving.

He waved his hand in front of the noseleaf. A burst of wavelengths appeared on the nearby screen. A beam of sound too high-pitched for human hearing had been broadcasting all the while.

He gestured for me to try it. I stretched out my hand and moved it back-and-forth. I expected to feel something, but of course the pulses of sound weren't going to register as tactile sensations. On the screen, however, the wavelengths shifted; they stretched out or squeezed together depending on how far my hand was from the source. Sutlive said the waves were bouncing off my hand and going straight back to the microphones in the ears. I could see the amplitude change as I pulled my hand further away from the emitter, the waves getting smaller. The long initial line, Sutlive said, came from the initial appearance of my hand—the smaller spikes were the echoes, traveling back. The shape of my hand, and its movement, were creating a pattern—a leaf would generate a different pattern, and so would the wingbeats of an escaping moth.

The hubbub around us suddenly seemed to subside. Sutlive checked his watch and turned off the batbot. It was now time for a critical design review: students in Müller's senior design seminar were redesigning the sonar head as a capstone project, and this was an early check-in on their progress. The deadline was a "product launch" at the end of the semester. It was a rare opportunity for an observer like me—typically, all the trial-and-error, brainstorming, revamping, and revising phases of development are revealed after the fact, after the end of a grant, after publication. An outsider sees the product perform after the curtain goes up, but not the evolutionary process of tinkering and selection that led there.

We joined Müller and half a dozen undergraduate engineering students around a seminar table in a small conference room with a digital projector illuminating the corner of one wall. This group was focused on refining the pneumatic system, the clear tubing I'd seen coiled around the Halloween bot's innards, intended to make the ears flap and bend with bursts of pressurized air.

"So, we are still going with the box?" Müller inquired, convening things without much preamble, taking a back seat, it seemed, in order to get the students to take the lead. There were nods of assent; the team members had divided up according to specific design tasks, but nobody had chosen to reimagine the body of the beast. I thought of Mazzolai's design of the appealingly tree-shaped Plantoid body as a nod to the audience, how it looked the part but didn't yet function like bark or cambium. The batbot didn't need to strike that chord for this audience. It could remain a box for now.

In robotics, I was beginning to realize, you can learn a lot from the afterthought, from the pieces assembled off the shelf. There are plug-and-play modules that, like some artificial version of convergent evolution, are already available to fill a need. If you want your batbot to fly, for example, there's a ready-made actuator for that, and a drone cage nearby for trying it out. Omar Khyam had already shown me footage of the bat head mounted on a commercial drone and buzzing around inside the mesh like an ungainly version of a box kite, something that really shouldn't fly. There's no off-the shelf actuator for bat ears, however—if you want it, you have to build it from scratch.

Sutlive began with a status report on the challenges of designing a pneumatic ear mechanism to replace the paddles that had been carved from wood. They'd created the mold for the design, but they kept getting bubbles in the silicone when they tried to print it. He passed around a work-in-progress—it was a tube of soft grey polymer, much smaller than the wooden paddles and divided into seven chambers. As the pressurized air flowed through these chambers, pressure would build up from one to the next, causing the actuator to fold over on itself like a crooked index finger. Placed behind the ear, the folding mechanism would function as an air-driven muscle, squeezing the ear into different shapes. A debate ensued over the advantages of two silicone polymers, both used to make zombie masks and other special effects in the film industry. One was apparently stiffer and easier for the 3D printer to mold without flaws, while the other was more flexible, more like the membrane of a bat's ear, but also harder to print successfully. Müller presided over the debate, nodding indulgently without intervening, until it was decided that they'd continue tinkering with the curing times and the molds.

One of the undergrad teams put up a series of animated CAD designs to display the new features of the system. The ear, for example, was a grayscale object crosshatched with lines as if it was made from graph paper—in simulation, it looked vaguely like a filter-feeding mollusk in the sway of a tidal current. One fundamental engineering goal, Müller observed, was to gain the advantage these tiny mammals have over current sonar systems by reducing the size of componentry—how could you make everything smaller and lighter, more parsimonious? Another was interchangeability: would it be best to integrate everything into a single ear design? Or could you make this design modular, so different components could be swapped out? If the actuators were small enough and modular in design, you could not only locate several of these actuators behind the ear as if they were multiple muscle groups, but also experiment with different arrangements, drawing from different shape formations in the ear zoo.

In response, the students said they were working on where to put sensors that would register changes in resistance and displacement as the ear shape changed, so the control system would know what the ears were doing. Sensors within sensors, in essence. One approach was to embed a thin nylon coil and a series of voltage detecting pins in the back of the ear. But embedding it would rule out a modular approach for the actuators—if you moved those around, then the sensor would need to be adjustable too. You could see the students working through these problems—not necessarily coming up with solutions but seeing the issues. At this pre-prototype stage, a change in concept could be made virtually, but decisions would have to be made.

Size was also an issue with the valves—they occupied a lot of real estate in the box that served as a body, and if the team included five actuators on each ear, then the pneumatic system would require a manifold with 24 valves. The valves would have to get smaller, Müller said, and weigh less—this creature was supposed to be airborne, after all. But they'd also have to be reliable. Once again, I thought of Semini's focus on developing bespoke valves—although in HyQ they were hydraulic, pumping fluid rather than air, they were nevertheless the key interface between the control system and the artificial muscles doing the actual work. The finer the calibration potential in the valves, the finer the gradation of movements that might be possible, and the closer to the twitch of an actual muscle. These valves would need to follow a similar path to become more bat-like: smaller, lighter, and capable of delivering ever smaller bursts of air.

The students filed out as another group arrived, tasked with a wider range of issues, including some of those who'd been testing things out in the lab. They began by passing around the mold for the next generation noseleaf, which at this point had already undergone generations of changes, to the extent that Müller labeled it bioinspired, an idealization of the original bat rather than an imitation. He held it up so he could see it from multiple angles, as if he was checking it against the features of bats he'd observed on some tropical expedition.

"This is the best noseleaf we've ever had," he said. "Very well done."

A sense of relief rippled through the room. Sutlive again mentioned the silicone issues in the 3D casting process. They weren't insurmountable, but there were still adhesions and bumps on the prototype. It was also true that the original specimen was essentially a death mask from the Smithsonian collection; Sutlive had smoothed out some of the bumps with the software but some distortion might remain. Müller didn't seem overly concerned.

"There's always some noise in the filter," he observed with a slight shrug. "Always a little vibrating going on." After all, there was some degree of variation at the level of individual bats too, the nicks and scars that come with living but also the errata of evolution, the spectrum of genetic variation among individual populations. In the end, what they were seeking wasn't an exact replica—as far as they could tell, leaving out some of the wrinkly bits in a bat's face didn't make any measurable difference in robotic performance.

Next up was the ear. A mold of the prototype was making its way around the table. This was the flexible structure that the pneumatic system would be pushing and deforming from behind. It would need to bend under pressure, then be rigid enough to return to its initial shape. The students put up an animated diagram of the shape, crosshatched with the mesh whose parameters could be lengthened or shortened to change the shape of the ear. Müller considered the prototype.

"This is more biomimetic," he said, running a finger over some of the bumps on the back where the soft pneumatic actuators would sit. "It looks really organic and natural. But we want to be halfway between here and engineering abstraction. Maybe we clean it up a little."

He pointed to the "skeleton" of the mold, particularly the part that ran vertically to the uppermost point—like the cartilaginous rim of a human ear, this "spine" would provide the structural support that would keep the ear from collapsing on itself.

"It's extreme," Müller said, thinking aloud about the prominence of the ridgelines that ran horizontally across the cup of the ear. "It doesn't occur like this in nature, but we see effects at the extreme range."

He cocked his head and pinched his finger toward his thumb.

"Maybe more tapering in the spine. It's a little too thick at the top."

The spine should be more natural, in other words. The rest of the ear, on the other hand, should be more stylized. It was a metaphor in the making.

As my plane taxied for take-off with the Blue Ridge mountains invisible beneath clouds that looked like bacon fat cooled in a skillet, I thought about how we'd navigate these conditions. As we burst through that cloud layer, we'd see almost nothing, but we'd be sending out and detecting position signals all the while, following our route over unseen landmarks until we reached our destination.

~

At this point, it felt a bit strange to have encountered a technological version of a bat, and even have a 3D black polymer facsimile of a horseshoe bat's pinna zipped into my bag with the pens and USB sticks, when I still hadn't seen the biological inspiration. It was a question of timing—I'd set up a visit with Susan Loeb, one of the nation's foremost experts on bat ecology who works with the Forest Service's Southern Research Station at Clemson University, in South Carolina's mountainous western corner, to coincide with an annual bat survey of Stumphouse Tunnel that takes place at the end of February.

The survey of this local cave is part of Loeb's ongoing research into bat population changes in a time of crisis; since 2006, the spread of a fungal skin infection known as white-nose syndrome (because of the characteristic white mottling of exposed skin on the face), has decimated some bat species and threatens others. Where it appears, vulnerable populations decline rapidly, often by 90%. Little brown bats, for example, have died by the millions over the last decade, and estimates of the total number of dead bats is approaching eight million. This menace exposed a dearth of data about bat populations—where did the 40 or so species found in the United States range? How common were they, and what was happening to their distribution and numbers as the fungus ravaged some species and left others untouched? To address these questions, Loeb was tasked with leading a team of experts to develop a set of standardized survey methods called the North American Bat Monitoring Program (NABat). The monitoring protocols are built around two main strategies: one is an acoustic survey, which might be accomplished by traveling a set route as night falls with a microphone on the roof of a vehicle, recording sonic data at certain frequencies, then running the results through a computerized recognition system that can pull out the distinctive vocalizations of different bat species. In this case, the car becomes a kind of highly sensitive, highly mobile ear, which is useful for species that roost in treetops rather than congregating in caves. It also means that the human driver never actually sees or hears the quarry. The other method is a colony count, which works for species that congregate in summer maternal groups and also for those that use "hibernacula" in the winter. This method creates an opportunity for a physical examination as well. After seeing the biosonar head, coming face-to-face with an actual bat was exactly what I was hoping for.

I met Loeb and a team of graduate students in a parking lot on the Clemson campus, which sits at the edge of the mountains in the western part of the state, surrounded by forest. Fluffy clouds were scudding across a blue sky—it seemed like spring, and indeed, although the trees were a leafless brown stubble along the crests of the hills there were daffodils nodding in the warm breeze, optimistic students in shorts, and ornamental plums already in full bloom. I wasn't sharing the springtime vibe, however. I was still in a state of flu-induced torpor. I felt like I could probably use a long session in a cave...of blankets.

Loeb was getting over a version of the affliction herself—I offered her a cough drop on our drive to the tunnel, but she had her own ready supply. She'd warned me not be fooled by the balmy conditions outside, but to expect a damp 52 degrees, with water drizzling from the ceiling and ankle-deep pools in stretches, which explained why we both looked like we were zipped up for a ski vacation.

The route took us up through the steep-walled valleys that characterize the southern Appalachians, where each small stream seems to have carved its own tiny cleft in the granite. Topographically, it's great for fostering diversity—for creating disconnected habitats where a global hotspot of salamander species abundance can evolve, for example—and dismal for trying to get around on wheels. That's why the tunnel project began—in 1830, exasperated railroad officials set out to dig their way under the mountains from here to Knoxville, Tennessee, connecting the port of Charleston with established markets further west. Two segments were chiseled from the granite over the course of several decades, but they never met: the costs proved exorbitant, the Civil War intervened, and 1,700 feet of tunnel was left for bats to use as a hibernaculum, and, for a time in the mid-1950s, for the university to use for curing blue cheese.

We parked below the mouth of the cave and began encasing ourselves in white Tyvek hazmat suits, which reduces the chance of transporting fungal spores between caves. Nobody knows for sure when the white nose fungus arrived from Europe, where bat species have evolved tolerance to it. It was first recognized in upstate New York, in the extensive complex of caverns beneath the little town of Howes Cave, which is kind of a mecca for spelunkers and day-trippers alike. I recall visiting the splendors of Howe Caverns myself as a child, although I can't imagine any bats roosting anywhere near the crowds I remember traipsing around down there among the mood lights and stalactites. In 2006, a caver photographed a strangely frosted bat clinging to the wall of one of these caves, the first sign of an emerging epidemic. The following year, hundreds of dead little brown bats were discovered in multiple caverns nearby, their emaciated corpses white with fungus. The strong suspicion is that the fungus hitchhiked on the footwear or clothing of cave visitors from abroad and subsisted on the walls of the cave until local bats arrived to hibernate. It has continued to spread across the continent, showing up suddenly in 2016 in a popular hiking area east of Seattle, over a 1,000 miles from the site in Nebraska that marked the previous limit of its westward spread. Testing has revealed the presence of the disease in several bat species around Seattle, and in multiple locations including Mount Rainer National Park.

I followed Loeb's lead—after sliding her feet into rubber knee-high boots, she tugged the elastic hood of the jumpsuit up so just her glasses and a fringe of curly silver hair showed on her forehead, then strapped a headlamp over the top and a hip pack around her waist. She was carrying a measuring wheel, in hi-vis orange, of the sort typically used by surveyors for marking property boundaries, and a clipboard. Everything we brought in, she said, would have to be wiped down with disinfectant when we emerged, including my phone.

We looked a bit odd, it's safe to say. Anybody can walk into the first section of the tunnel, and those who showed up regarded us with wide-eyed speculation. One car actually did a full 360 and came back, the driver poking his head out the window to ask what was going on. I could see what he was thinking. Murder investigation? Radioactive contamination? Alien crash site? When he heard the actual reason—bat research—he seemed disappointed.

The mouth of the tunnel had a hand-hewn quality to it, as if each angle of the rounded arch that stretched 25 feet high above us had been hammered into the glistening horizontal ledge by an army of miners. It was indeed a watery place. And surprisingly loud. One stream was cascading over rocks to the side of the opening, and another was trickling out from the entrance itself. Boughs of waxy green rhododendron leaves hung down from above, forming a tunnel of their own and lending the scene an almost tropical air. Passing the threshold, which was coated in vivid green moss, I could feel the cold damp breath of the mountain's interior wafting across my face, could feel the rampant mustiness in my stuffy nostrils even if I couldn't smell anything. Just inside, a curtain of water pinged off my hood, slow and incessant, with the thrum of a just finished shower. A cold shower.

The team was already eagerly scanning the walls with handheld spotlights, spreading out, the beams arcing across the jagged ceiling, pausing over crevices in the walls. I turned on my own light. The walls were rough, and there were lots of shadows that looked like splayed bat wings to me. I thought of a story Rolf Müller had told me, about coming across what is often called the world's smallest mammal, the bumblebee bat, *Craseonycteris thonglongyai,* while looking for other species in a cave. In the flashlight beam he could see these tiny shapes clinging to the walls—they looked exactly like bats, he said, but at first he couldn't believe that's what he was seeing, because they were only about the size of a fingertip. For him, it was a matter of adjusting the search image, of recognizing what was there. I had no search image, no size in mind. Every protrusion, every mineral deposit, looked like a bat to me.

One of the searchers had stopped ahead, her beam frozen on what appeared to be just another barren gray patch of ceiling.

"I have one here."

The news rippled through the group, and we gathered eagerly around the spotter's outstretched arm, peering up to where the white circle of light fixed on what looked like a small rusty lump about the size of a kiwi fruit.

Loeb had warned me on the drive over that we might not find many bats—or, worst case, none. As recently as 2014, when the disease arrived in the cave, over 320 bats were found by the survey, and the count would've taken most of the day. Last year there were only 31, although the numbers appear to have stabilized at around 30 after the initial steep decline. Mostly, these were tricolored bats, *Perimyotis subflavus*, one of nine species in South Carolina known to utilize cavities or caves, and thus one of the most vulnerable to disease. Five other species roost singly in trees, which minimizes the chance of exposure.

Tricolored bats get their name from the way each hair changes color from dark brown to rusty brown to yellow from base to tip, although that wasn't visible from down below. Nor was there any discernable head or tail; at this distance, what seemed to distinguish bat from stain or tidbit of quartz were the protruding knobby bones of the wings, which bracketed the body vertically like two sides of a rectangle. I had my search image.

Loeb brought up the rear, rolling the measuring wheel along the gravel floor of the cave. While she did so, another assistant aimed a red laser dot at the rock surface near the bat to take an infrared temperature reading. He whispered the result to Loeb, who marked it on her clipboard. That data is key to understanding why this epidemic is happening. The bats seek out the right spot to roost partly according to temperature, hoarding calories as they await the spring emergence of insects, going through a cycle in which they slow down their metabolism (and their immune system) for several days before rousing themselves for a period of about 15 minutes, then returning to slumber.

The problem is that *Pseudogymnoascus destructans* thrives at certain temperatures too. Temperatures above 70 degrees inhibit its growth, as does sunlight, but in constant cool temperatures in the mid-fifties, like those found in the cave, it thrives. The fungus doesn't need to spread from bat to bat; if it did, then tricolored bats, which are frequently solitary in hibernation and raise their young in the tree canopy rather than caves, might not be so vulnerable. Instead, spores can shift from infecting the living to infecting the dead, colonizing the decaying organic material that might come to rest on the cave floor, like dead bats or insects, scavenging enough nutrition to mature and release more spores. All of which means the bats, in seeking a climate-controlled haven, are actually exposing themselves to the right conditions for infection.

We found several more *Perimyotis subflavus* in the first section of the cave, including one that was just tantalizingly beyond the outstretched latex fingertips of the team's tallest member, who explained the survey protocol in a low voice: in order to conduct a physical examination, the bat has to be within reach from the cave floor, no jumping allowed. The deeper we got, the more the light receded in the distance like a clerestory window or some celestial object, and the more fully the survey routine took hold. Everyone grew quiet, pointing where possible, flicking their headlamps over to the red light setting, trying not to disturb the bats unnecessarily. Unfortunately, I kept forgetting to turn off my headlamp beam as I observed not just the bats but the observers, which resulted in a lot of half-blind pointing in my direction.

Eventually, we reached the brick wall that marked the end of the public section of the tunnel. A panel of fortified rusty screening blocked the way like a portcullis, lending an almost medieval, *Lord of the Rings* tenor to the process of finding the key, unlocking the hefty padlock that secured the heavy chain and stepping one-by-one into the beyond. The graffiti was heavier here—the walls of this notorious party spot were layered with spray paint mementos of love and immortality, some of them decades old. Above the door was the empty socket of what I assumed was once an outdoor light fixture, with what appeared to be a torpid bat hanging from a hole in the galvanized metal.

"That's a dead one," one of the search team whispered as we waited to pass through. "See the bone in its wing?"

I wasn't alert to the difference between torpor and death, but now that he mentioned it, I could see that one of the wings was splayed out rather than tucked in like the others, and it didn't appear to have much in the way of wing membrane attached. Seeing a mummified corpse like this is unusual; normally, a dead bat, whether a victim of white nose or some other cause, would simply drop to the cave floor and decompose.

On the other side of the door the water got deeper—I could hear the team sloshing through it as they headed toward a mound of startlingly green broken pine branches illuminated from far above—we'd reached the air shaft, although water was also using it as a conduit, and the whole scene was bathed in the ethereal spray of a waterfall. The floor of the cave was now a long pool layered with the woody remains of vegetation that had fallen in landslides from the surface, so that, with each step, we sunk through a lattice of dead limbs, like crossing a beaver dam. I wanted to get through there quickly, not just because a spring torrent was splashing on my head, but because I pictured something else coming down through the void at any moment, like the enormous spider I saw spread like a handprint on a pale smooth rock face on the other side of the air shaft. This was not the place for phobias.

Not far along, the spotlights converged on a low section of wall. We'd found the first bat within reach. The researcher who found it waited for Loeb to catch up, increasing the suspense. Then, with her gloved hand dramatically illuminated, she reached up into the center of the light and plucked the bat gingerly from the wall, turning it face up as she did so. We circled around her, readying ourselves for the procedure: it reminded me of an operating room, the tools set out just so, the surgical team in scrubs surrounding the illuminated patient in an otherwise darkened room.

The bat's tiny face protruded from a mantle of white latex—finally, I was face-to-face with an individual biological specimen, rather than just *a bat*. It was a moment that clearly meant a lot more to me; the bat regarded us all with what I took to be a peevish WTF expression, nipping the gloved fingers that held it to no avail. I was relieved; I half-expected its face to be plastered white with the fungus, but it appeared to be growth-free.

I thought of the nose and ear zoo I'd seen in Blacksburg, and the polymer bat ears I'd seen. This was no horseshoe bat: the nose was more like a half-opened bud than a leaf, just a tiny skin tag that the shaggy fur made almost invisible. This species tends to hunt in the open, over water or between large trees rather than swooping into the undergrowth, and its ears were like leathery kidney beans.

What was most notable was the protruding furry mound of the brow, which seemed to squeeze the face into a sliver of real estate, where an expansive mouth was working frantically, exposing twin rows of glistening scalpels. It looked like the bat was screaming, or snarling, yet almost no audible sound emerged, at least at wavelengths I could detect over the watery drips and drabs in the background. It was only when I played back the video I'd taken with my phone that I heard it clearly, suddenly amplified; the microphone had picked up a raspy, staticky buzz, like a wind-up toy, only faster. The bat was not only protesting but trying to figure us out as well, enveloping us in acoustic waves, sensing us with sound.

"Poor guy," someone said, after various size measurements of wings and abdomen had been taken with a finely gradated blue ruler, and a tiny metal band had been clipped to one of its wing bones. They'd spotted the fungus, which looked like a crust of chalk, not on the nose but in the creases of skin near his genitals, and then where the wings met the body, in the armpit, if a bat could be said to have such a thing. The white nose fungus is keratinophilic—it feeds by breaking down the keratin proteins in a bat's skin and hair. There are several theories about why a skin infection would be fatal in bats—after all, the athlete's foot fungus causes an itchy and annoying infection in humans, but it doesn't kill us. One hypothesis is that the itchiness drives bats so crazy that it wakes them up, which might explain why diseased bats have been observed flying around in the middle of winter. Recent studies comparing disease-free bats with infected ones show a striking disparity—at the end of a fasting period, the infected little brown bats had far less fat than their counterparts. They'd burned through their reserves twice as fast as the control specimens, and if the infection had reached the wings then there were differences in their overall body chemistry as well, suggesting a systemic stress response.

The researcher holding the bat spread each of the wings wide, smoothing them out with a gloved fingertip, then stretching them between two pinched fingertips so this extraordinary structure was visible. Bat wings are, in principle, akin to our upper extremities—they resemble hands that have been stretched out to three times the length of the body, with webbing between the splayed fingers, as if taking a cue from the ribs of a maple leaf. Rolf Müller had said that, from an engineering perspective, the flexibility of the bat wing was another unique feature worthy of emulation; while the bones of birds are fused together in the wing, in bats each of the elongated finger bones can adjust independently in flight, changing the contours of the wings across multiple angles and planes, matching the precise triangulations of the biosonar to the exactitude of the flying apparatus. The scrim of flesh between our human fingers offers no comparison really; trying to imagine it, I had wiggled my fingers, but there was obviously no air resistance between them. The best comparison I could come up with was the effect of wearing a pair of swimming paddles in the pool, or playing the guitar, the finger contortions each chord change requires when a maestro like Leo Kottke takes flight.

Often in a more advanced stage, the infection will cause lesions that erode away the membrane, pocking the wings with holes, but this guy's wings had no visible pinpricks, which was a hopeful sign.

One of the other researchers flicked on a UV wand, which glowed like a miniature light saber as they moved it slowly, almost ceremoniously, in a first pass over the bat's wings. It was a nonverbal signal to everyone else to turn off their headlamps.

"Your light," someone whispered.

"Huh?"

"Your light is on."

Once I got with the program, we were all cloaked in darkness save for the purple glow of the wand. It picked up a faint coating, like talcum powder, between the bones.

"Poor guy," someone repeated.

"The wings heal quickly," Loeb told me as she wrote down his vital signs. "He might make it—it's late enough. But there's also an immune cascade that happens, so..."

When the bat's systems go back into active mode in the spring, the immune system suddenly starts fighting the infection. The immune response itself can kill an already weakened bat.

We weighed the guy in a tiny mesh bag that dangled from a miniature hanging scale—a typical tricolored bat weighs less than a quarter of an ounce, so it barely depressed the gauge. Then the searcher who had found him carried him back to the wall, propping him up until he re-attached with his claws. We left him in the darkness.

We found several more tricolored bats at eye level as we headed toward the back of the cave where the temperatures and humidity of the lower rock faces were apparently more to the bats' liking. Several were coated in droplets of condensation, so they gleamed silver in the glare of the headlamps. We ducked through the doorway of another bricked wall and into the last chamber—it was covered in graffiti, especially the cave-within-a-cave carved high into the rock, as if the workers had stopped abruptly one day and never went back. It looked a bit like a desecrated altar. We'd reached the end.

Outside, after we meticulously disinfected our boots and disencumbered ourselves from the Tyvek suits, and I'd disgorged some tissues from an interior pocket and finally blown my nose, Loeb tallied the numbers. All told, we'd found 34 bats, about the same as last year.

It was a relief, but I still wondered what it would've been like to see hundreds more in scattered constellations along the walls. It's happening with amphibians as well, this dramatic disappearance, this reversal of former abundance. What does biomimetics offer in reply to such calamities? Directly, not much. However, you can't be tuning in this deeply to the frequency of bats and not witness what is happening to them. I thought of Rolf Müller's work with the Smithsonian, cataloging the physical features of dead specimens. You can learn a lot from that kind of collection. But I also recalled his story of being face-to-face with a live bat and realizing that the facial features of this creature were in motion, were twitching and vibrating, too fast for the naked human eye. That was the moment that inspired the research question, that led to the collections in the nose and ear zoo, and it's that kind of moment we'll lose if these species disappear. As a question, what is *was* like to be a bat offers far fewer possibilities than what it *is* like to be one.

~

Two evenings later, I was out late in our garden in Columbia, raking up dead oak leaves from where they'd collected under the azaleas, exposing the soil where the early spring bulbs were poking up their blunt green heads. The robins had been coming through in waves and pausing to feast on cherry laurel fruit that had mellowed over the winter, and that evening they were roosting quietly but still flensing away the black flesh, so that there was a constant patter as they dropped the pits through the leaves, like the first big drops from a rain shower. It was chilly, but calm.

I looked up and there, circling the sky, was a bat. Yes, *a* bat. Occasionally its body caught the traces of sundown, and I could see it was brown, but that was about it. One faceless little brown job zigging and zagging across the darkening sky. I had no idea what it could possibly be finding to eat up there; one of the reasons I was out raking at this hour was because there were no mosquitoes yet. I wondered why it was alone; solitary behavior now seemed like a sign of scarcity and disease rather than just the norm for certain species. Although fluttering around in an aerial death spiral is a hallmark of white nose infection, Loeb had said that some species, including tricolored bats, will emerge from torpor in early spring and make the occasional foraging run. She'd said there's also early evidence that some bat species may actually be expanding their range as those devastated by white nose syndrome disappear. The eastern red bat, the species that hibernates among the leaf litter, might be one such candidate. I had to hope that a single bat in an empty sky was not an omen.

~

There's another way to approach this question of what it's like to be a bat, and it involves what the sound artist Alex De Little calls "sonic empathy," the creation of experiences that make us focus on the act of listening, and how another species might experience sound. De Little, an experimental musician who just completed a doctoral degree at the University of Leeds in the nascent field of Sound Studies, had contacted Rolf Müller with an idea for an interactive installation at the National Science and Media Museum. He would create a pair of bat ears, based on the models in the ear zoo, and then invite participants to wear them. The project would be titled: "What Is It Like to Be a Bat?"

Müller sent the specifications for a bat model with appropriately extravagant ears drawn from the Smithsonian Collection. *Plecotus auratus*, the brown long-eared bat. It's a common species in the United Kingdom, where it gleans insects from leaves along forest edges, but it's mostly distinguished by its enormous pointy ears which splay upward and outward like a pair of windsurfing sails. They resemble the greater horseshoe bat pinnae the lab was using as a model for the batbots in Blacksburg, but these are proportionally much larger, almost as long as the four-inch bat's body. These giant sound catchers were what De Little and an engineering colleague, Nick Fry, used to create an imitation in wearable form.

I met DeLittle at his studio in a gentrifying district of Leeds; the building, still under renovation, is a former convent hard by the junction of several roaring highways and elevated onramps. To get there, I wandered up from the downtown train station in the summer's record-breaking heat, past a small encampment of tents perched on an old hillside green with a view of rectilinear cranes framing the city skyline.

It was loud. Really loud. Buses hissed. Trucks rumbled. Silence seemed impossible. One of the reasons the nuns left, De Little said as he gave me a tour of the different floors, was because they got tired of all the noise. Yet he had the big windows to his studio open, with a view out over the conjoining of dual carriageways and the road noise reverberating off the walls.

His work is based on a practice known as Deep Listening, developed by the avant-garde composer and music teacher Pauline Oliveros to encourage active engagement with the soundscape that surrounds and permeates our bodies. "I'm interested in doing listening," he'd told me on the phone, in creating "secular guided meditations that dialogue with space." He mentioned one of Oliveros's best known compositions, a set of meditation instructions that includes an invitation to walk around barefoot at night, listening in the dark as if the soles of the feet were ears. De Little also described his work in terms of the emerging politics of Sound Studies, the recognition of an acoustic epistemology or "acoustemology" which challenges our "ocularcentrism" by identifying an alternate, and potentially subversive channel of experience. It all sounded to me like part of the same expanding trajectory of ideas about the status of what is out there and how we know and experience it, this world of animate objects and non-human agents with which I was becoming more familiar, if still not altogether comfortable.

The bioinspired bat ears were perched on the head of a smooth black plastic mannequin whose only feature was a slim, Modigliani-like nose. The ears were also black, and they were so large they looked more like a ceremonial wing sculpture or headdress than ears, although De Little said they had to scale them down from the actual bat proportions to make them wearable at all—the real ones are so outsized that the bat tucks them down under its wings when it roosts, like tent flaps. To be biologically accurate they would have had to be several feet in length, not 18 inches.

Lifting them with two hands because they were surprisingly hefty, I tried them on. They'd been printed in sections with a 3D printer and then glued together, so up close you could see the seams forming angular polygons that crosshatched the more subtle lobes and curves. The rigid polymer they'd used was not unlike the black model I'd been given by Joseph Sutlive in Blacksburg, which I still had in my bag as a kind of keepsake. Along the outside edge was the smooth ridge of the helix, cupping the dish of the interior where another ridge, the anti-helix formed. Near the base was another ripple in the polymer representing the targus, the nub of cartilage we have right where the auditory canal opens, and which in bats is much more membranous and pronounced. In *Plecotus auratus* it's especially prominent, almost like another pointy ear jutting up to fill the cup of its bigger neighbor.

There was a strap around the back, but to keep them over my own ears I had to hold them up with my hands, elbows splayed out. The plastic pressed my ears to my skull, which was uncomfortable but part of the design: originally, De Little thought they'd make it so the human ear was pushed outward to amplify sound, but Müller had told them to pin the ear back to expose the auditory canal.

The effect on my hearing was at first not so much augmented as muffled; I couldn't hear De Little's voice very well. Then he invited me to point my bat ears toward the windows and turn my head slowly. He watched me with a portentous half-smile as I rotated. I wondered what expression was on my face—here I was, turning in a slow circle with two giant plastic extremities protruding from my head. There was no mirror in the room—for this I was grateful. As I turned, I noticed something change, and De Little's face noted the shift—he saw me get it. At a certain point, the sounds of tire and brake and tarmac reverberation were suddenly crisp, each one distinctive, as if the full spectrum of roadside experience was being poured into my skull. Then a slight shift in the angle of the ear... muffled again. You could see why ear rotation would be a necessity for bats on the wing. I imagined all the possibilities in the ear zoo, all the different shapes, the different acoustic experiences each would offer.

To deepen the experience, De Little asked me to try an exercise. Spin around with your eyes closed, he said, while I snap my fingers. This was something he liked to do—put blindfolds on people, ask them to shut their eyes, make them face a corner, anything to deprioritize the visual and make people concentrate on the sounds they were experiencing. Now not only was I turning in circles with big plastic ears on my head, but I was blind too. *This feels a little weird*, I thought as I swayed to the rhythm of the sharp snaps from De Little's practiced fingers. *This is what HyQ might have felt, if a robot had feelings. Trotting along blind, stumbling over bricks...* I could localize the sound but also sense how it changed as my angle changed. Also, I could feel my eyeballs trying to travel to the sound, trying to find the source visually even with my eyes shut. Was I like a bat? Not perfectly or deeply. But I wasn't conventionally human either. I was attuned to the sound of being self-consciously human. Here was another phenomenology, another way into the experience of being in the world. It wasn't a nature walk or an exercise in ontography or metaphorism. It wasn't visual. The closest I'd come to it was in the company of an accomplished birder who rarely lifted his binoculars, noting, with a raised finger and unfocused eyes, the percussive signature of a woodpecker drumming way off in the woods.

After I removed the ears, which also seemed to amplify the heat, we continued with this sound practice out in the hall. De Little's musical scores, which are collected in a limited-edition book he gave me called *Spatial Listening*, are similar to Oliveros's early work: as with a yoga or meditation workshop, these are instructions for participants to perform in specified settings, which he describes as "human algorithms." There are "resonances" and "reverberations" and "echoes" to explore. It hadn't occurred to me that there was any difference between these vibratory phenomena, but then I'm a bit sonically challenged. I'm a visual person, primarily. I sketch, I doodle with music playing in the background, but I don't play any instruments.

De Little took me out into the stairwell where we stood on different levels so we couldn't see each other. He clapped, a single percussive note, and I waited. The goal was to wait until silence returned, until I perceived the reverberations had dissipated, and then to respond with my own clap. The sound of his hands striking each other resounded upward in this vertically open space, vibrating the air in what almost felt like rings growing wider and wider after a rock splashes into a pond. I could feel the sound waves as they died out, not all at once, but as a tonal sequence, which De Little had described as an interaction with the space. I wondered if we might think of this sonic architecture experience in terms of "intra-action," the term coined by Karen Barad for these experiences of connection between different living or nonliving agents, between my body and the walled hollow of the space, for example. Individual agents like the convent and I, she has said, "only exist within phenomena," defined by the mutual act of relating to each other. Deep listening draws our attention to the sound that bounces back and forth between us, defining this relationship in a particular moment that lasts until the clap reverberations disappear. De Little has described the practice in similar terms: "By sounding, you act on the space, and in return it acts on you. As you stretch your ear towards these sonic essences you create the architectural environment in your listening, but you also constitute your auditory self, here, now, in this space."

To explore resonance, we took our auditory selves to an empty room with peeling floral wallpaper and the ghostly impression of nuns taking their tea. De Little stood in the corner with his face touching the walls. He began to make tonal sounds—deep and throaty at first, then rising through the register toward a sustained yelp—it sounded like someone moaning while they struggled with a dream. At certain points in this escalation of sound, he said, the reverberation begins to match the emanation, creating that effect you get with a prayer bowl, a kind of hovering around the fovea, as if the walls are moaning back.

We tried a smaller room, once a sitting room perhaps, or a bedroom.

"Ready to try it?" De Little said with a puckish grin, as if he could tell that I only ever sing in the shower, and nowhere else. I put my nose to the musty wallpaper, wondering if I had ever been naughty enough to be punished this way in school.

"Close your eyes," De Little said.

I started with a tentative but chesty growl, which came out like the chanting of Tibetan monks…if someone had instructed them to do it while whispering. The sound was flatter somehow, less rich since it had less room to reverberate. The corner concentrated the soundwaves, so they struck the wall in front of my face and immediately rebounded, which became even more clear when the tenor of my voice matched some acoustic propensity in the room. It was an eerie sensation, feeling the vibration of my own moaning, as if my head was at the center of this single resonating note. It was almost the antithesis of echolocation; the sound felt as if it was everywhere at once.

After resonating, we went outside to the street to practice echoes, equipped with wooden percussion blocks and rubber mallets. We stood opposite a brick building on the other side of a parking lot, held out the block and smacked it with the mallet. It made a dull *tock*. The sound was supposedly bouncing off the wall of the building and coming back to us as an echo, but I could barely hear it. I asked what the difference between an echo and reverb was—aren't they both sound waves bouncing off something and coming back? De Little said an echo happens in uncontained space—it dissipates everywhere but the surface it strikes, whereas reverberation happens in enclosed spaces, where it can bounce back and forth off multiple surfaces.

The difference became clearer as we traipsed down through the park to a stand of enormous ash and sycamore trees, hitting our percussion blocks the whole way. At first the tocks disappeared down the hill into the downtown skyline. Huddled near the tents, a trio of smokers observed us—and heard us presumably—exploring our aural selves.

*Tock…tock…tock…*As we got closer, De Little, who didn't seem at all fazed by being out in the open making sounds without any purpose other than to listen, looked at my face to see if I'd noticed. I had—the *tock* was bouncing back from the direction of the tree canopy, louder than the brick wall, with just a slight delay between the mallet strike and the echo coming back.

"What you just experienced is like what Rolf was doing with the trees," De Little said, recalling the alfresco batbot explorations of Virginia Tech's tree-lined campus. I knew where the tree canopy was because I'd heard the echoes—I could've pointed to it with my eyes closed. We were experiencing the acoustic landscape and constituting our aural selves in response.

But not like a bat, exactly. Like a batbot.

3

INTERPRETIVE STATION

Social insects and robotic swarms

Recently, a colleague had an unusual problem: unbeknownst to her, a colony of feral honeybees had taken over the wall in her bedroom, right above the headboard of the bed she and her partner shared. They'd hired someone to paint the exterior trim boards where the roof line met the brick wall of their house, and the painter returned from an initial inspection with reservations—I'm not doing this, he said, until you take care of those bees.

You can imagine how they felt.

Bees? What bees?

He took them around the house and pointed to where the brick façade met the white fascia boards that ran horizontally under the eaves. From the outside, it looked like just a small cluster of honeybees had stopped to gossip while out running errands, right at the line where fascia meets brick. It was late spring; the azaleas and magnolias were still in bloom, the air heavy with the scent of flowers, tinged yellow with pine pollen. Good times for feral bees.

Instead of freaking out, as I almost certainly would've done, my colleague called a local apiarist she found online. Many beekeepers love feral bees, because of the genetics—despite all the challenges domestic bees are facing in this country, from mites to disease to pesticides, wild bees seem to thrive.

Feel the wall, the husband and wife team of bee enthusiasts said, when they arrived for an inspection. Press your palm against the wall.

It was warm. Warm with the collective endeavors of thousands of insects, working together behind the drywall to build this parallel nonhuman abode.

I was there when the beekeepers began the extraction process, cutting through the drywall to reveal long pale lobes of hive that looked more anatomical than insectoid, like lungs or glands suspended in the body cavity of the house. They were punctured in places from the initial incision, oozing, encrusted with a dark simmering layer of agitated worker bees.

The removers, Bill and Ginger, were dressed in baggy white tunics that zipped up into an oval mesh head net, with thick gloves covering their forearms and half a roll of duct tape wrapped around their ankles. It made them look a bit like they'd dressed as extraterrestrials for a costume party. Bill was using a putty knife and flashlight to gingerly pry away the comb, handing it down in dripping sections to Ginger, who pressed the chunks into wooden frames strung with what looked like dental floss. She slid the frames down, like hanging files, into a wooden box known as a Langstroth super—a bee hive in the making. The hexagonal cells were full of pale honey—nectar from the plants outside, essentially, since the bees hadn't gotten around to fully masticating and fermenting it yet—and uncapped brood, row after row of pearly larvae. The floor was sticky with honey, and stray bees were silhouetted against the window glass and buzzing inside the opaque globe of the ceiling light. We swiped a taste of honey from a stray chunk of wax—it reminded me of the hint of sweetness you get if you pluck a plume from a honeysuckle or clover blossom and nibble the white end, floral and sweet.

It wasn't something you could savor, however, because the entire room was vibrating with a sonic version of electricity, as if a powerline had gone down and was crackling with voltage on the sidewalk. This is what it's like to be an outsider on the inside of the swarm. You have the sense that what's going on isn't random; there are rules of behavior here, distributions of labor, priorities being assessed and acted upon. The chaos is coming from your presence, from your interruption of the routines of the hive. The bees are civilization; you're the storm.

Bill was listening attentively. He kept reaching up with the end of a customized shop vac (lots of duct tape), suctioning hordes of *Apis mellifera* away from the comb, and then turning off the roaring machine to listen. Was he hearing a change in pitch? A hive deprived of its queen is said to roar like an engine revving at high speed; Bill said he was listening instead for the high C note that workers produce when the queen is present. The flightless, larger queen was the prize here—find her and the successful hive could be moved more or less intact; lose her, or kill her by accident, and the hive would have to be requeened, with no guarantee that the replacement would be as successful.

He wasn't hearing it.

"Boy, they've been in here for a while," he marveled, prying away yet another layer to reveal thousands more bees and darker layers of comb. "This is from two years ago."

He cut away more drywall, a rectangle that reached to the crown molding. The comb was spreading horizontally, but also vertically, heading for the ceiling. And beyond that...how far?

No single bee has the blueprint for where the hive expansion will go, but the decision still has to be made. The evolutionary biologist Stephen Pratt, writing about the way honeybees come to collective decisions about hive construction, suggests colonies have to confront such questions regularly, not just how much honeycomb to build, but also how they time their construction. The stakes are high: spend too much energy building comb instead of making honey when nectar is abundant, and the hive will starve in the winter. Don't build enough comb, and there won't be enough room to store honey efficiently when there's plenty of nectar around. Once again, the hive will starve. On top of that, there's the question of how colonies "regulate the *type* of comb built," deciding whether to build small honeycomb cells for workers or large cells to produce drones, the male bees who don't do anything much at all unless there's mating to be done. More drone cell construction means the bees have decided the hive is doing well enough to plan a split and start a new colony, since having drones around normally only matters when a new queen departs the hive to found a new colony. Choosing to build more worker cells, on the other hand, means the hive is probably understaffed and playing catch up with the chores.

Bill peered into the shadows with his flashlight beam. The homeowners hovered in the background, amazed to a degree, but mostly dismayed. They had their own set of questions. How much of their house was going to be dismantled? How had they ever slept here with all this industry going on behind the pillows?

The other distinctive element of this soundscape was some relatively innocuous, Bible Belt cursing. You might think that beekeepers get stung so much that they must be inured to it—maybe they don't even feel it anymore. But the toxins in a bee sting are designed to ensure that doesn't happen.

"They are not happy right now," Bill observed drily, with the restraint you might use if you were afraid any loud noise might cause a bomb to go off.

"Argghhh...you son of a...*gun*."

A mad scrabbling at shoulder blades, a mini waggle dance on the step ladder. The bees weren't bothering me; nor were they interested in Ginger. Instead, the defenders of the hive were crawling in slow motion along the moon surface of Bill's back, prospecting for the spot of their fellow's sting so they could sting some more. In addition to sound, bees signal to each other via scent—the smell of a disturbed hive is said to be reminiscent of bananas. Once the stinger of one worker embeds itself in your flesh, the gland at its base continues to pump out mini-clouds of a pheromone signal to the other bees, something along the lines of "Kill! Kill! Kill!"

"Should've worn the other jacket!" Bill complained, grimacing. "It's too tight across my back!"

Ginger caught the falsetto note in his voice and looked up from putting a lid on yet another super, eyebrows raised.

"Do you need me to get your epi pen?"

His epi pen?

No, not now, Bill said gruffly, and went outside for a breather.

There were more stings, and more curses, before the layer from four summers ago, ambered with age like an old hardwood floor, was stowed away in one of the supers and loaded into the truck with the shop vac full of bees. The queen, they hoped, was somewhere inside.

~

Bees and ants. And locusts. And schools of fish. Lines of migrating geese, wings almost touching, and prides of lions on the hunt. Flocks and swarms of these creatures all share certain properties. Take, for example, Pratt's analysis of honeycomb construction. Deciding how much comb to build means "poorly-informed" worker bees must "generate a pattern attuned to global colony and environmental conditions even though they typically have direct access only to very limited local information." Under these conditions, in other words, there's no central command; no single honeybee decides how many drone cells to build, just as no single bird dictates the astonishing synchronicity of starlings as a murmuration whirls through the sky. Decision-making is distributed throughout the members of the group. Everyone operates according to basic rules of behavior and the pursuit of a shared goal, but they act locally, on an individual basis, adjusting their own behavior as they receive new information from other members of the group and the environment. Complex behavior like comb building emerges from this collective intelligence, this "hive mind" gathering, communicating and adjusting to new information.

It's the tilt of a wing or the flick of a fin, multiplied a thousand-fold. Or, perhaps, the repositioning of antennae and wheels as swarm robots interact and adjust...spontaneously, tumultuously, and yet, logically. So far on this journey, the robots we've encountered have been solitary in nature, the complexity of their achievements measured by an individual yardstick. HyQ, for example, was a creature of great individual complexity; the goal was to mimic all the properties that allow locomotion in an individual animal. Swarm robots, on the other hand, are more like ants, building morphogenic bridges of living bodies over water, or feral honeybees constructing elaborate dwellings in the wall of a house—the compelling questions they raise come from their collective behavior, from the ability of the group to accomplish things that no individual could achieve.

As I stood on the periphery of the *Apis* swarm, did I think of drones swirling in coordinated sorties around their charging station? Not immediately. It's hard to think in terms of analogies when you're tingling with the vague sense that an assailant is crawling between your shoulder blades. Once I got outside, however, and sealed myself in the ambient hush of the car, I started thinking about the natural history of social insects, and wildness.

What is wildness? Often, as Gary Snyder has observed, wildness is defined as what it is not—not civilized, not domesticated, not tame. Instead, Snyder offers a series of defining characteristics that emphasize freedom and spontaneity; animals, for example, are "free agents" whose behavior includes "fiercely resisting any oppression, confinement, or exploitation," while societies are those whose "order has grown from within and is maintained by the force of consensus and custom rather than explicit legislation." What might look disorganized and chaotic to a cursory observer is actually a kind of idiosyncratic order, operating according to local rules. An "ordering of impermanence" perhaps, to put it in Snyder's terms for the "process and essence of Nature."

Now it's certainly not Snyder's purpose to promote consideration of the nature of machines here; after all, he encourages shooting an arrow into the heart of the "occidental scientist-engineer-ruler" beast to slow down technological progress. It's probably quite heretical, in fact, to invoke him here, but that's exactly what I propose to do. Let's consider the possibility that our technology might lie on the same wild path. Let's consider these other "free agents," whose collective networks emerge from within and are maintained by the force of consensus rather than explicit legislation from some central governing authority: robotic creatures in the swarm.

What is wildness in a machine?

~

Our journey is going to take a myrmecological turn here, down an ant trail of sorts. The biomimetic connection between ants and robotics is longstanding; before swarm robotics really took off in a kind of Anthropocene explosion of biomimetic diversity, groups of rudimentary agents interacting via behavioral algorithms were known simply as ant robots. Ants make useful metaphors for the way we imagine machines, partly because we recognize a kind of kinship in the organizational challenges we face as social species. There's clearly something about ant society that lends itself to anthropomorphic metaphors, to "royal castes" of queens and courtiers marching down "streets" and "highways." The famed ant biologists E.O. Wilson and Bert Hölldobler have gone even further, pointing to a "strange parity" between our species—we're roughly the same in terms of planetary biomass, and ants, like us, tend to "dominate" the ecological landscape, pushing other less social insect species to the margins.

At the same time, it's an odd equivalency, because ants are so different from us in the premium they place on the system over the individual. Not only are worker ants rolled out in the biotic equivalent of a factory floor, meant to be expendable and operate with very limited instructions, but in what seems like the script from a dystopian novel, an ant's ability to reproduce is largely dictated by the station to which they are born. The defining feature of the "eusocial" insects, what distinguishes them from other insects that live in more loosely organized colonies, and even from highly social creatures like us, is that in most species, workers are programmed to forego their own reproduction and serve the needs of a reproductive caste, composed of a limited number of their own relatives. Under normal circumstances, workers do not compete for mates. Nor do they raise their own young. There's no romance in the life of your average ant, no individual quest for self-fulfillment and meaning. Paragons of nonhuman domestication, they essentially work themselves to death. It's self-abnegation taken to such extremes that Wilson and Hölldobler have likened the ant colony to a "superorganism," in which individual ants function like cells in the body of the nest, following biochemical cues without worrying about their own survival or reproduction.

At first glance, none of this sounds much like Snyder's definition of wildness. In fact, it sounds like a system designed to squelch even the faintest hint of spontaneous self-expression from ever breaking out, the apotheosis of the tame. Yes, individual ants have the power to make decisions and take action without asking permission from some central authority. (As I write this, ironically, an Argentine ant has begun scurrying across the screen of my laptop, as if to emphasize her freedom to roam.) And yet, the behavioral programming is so rigorously codified in each ant that collective projects normally proceed in a logical, albeit randomly determined manner. The behavioral rules encompass and exploit randomness. Serendipity is actually part of the system.

Could it be that another species is more civilized than we are? Does that make them less wild than us? It seems I usually conflate otherness and wildness—the wild is always other species; we impose order on their unpredictability; we tame them by making them play by our rules. What ants do challenges those assumptions. Whatever wildness we're going to find isn't going to be the kind we define in opposition to our own culture and civilization. Instead, it will be an order of its own, its symmetry arising paradoxically from chaos, rising and falling on its own time scale. This central irony, that "crazy ants" aren't in fact so crazy after all, that what appears to be uncivilized and disorderly conduct is in fact quite systematic, is the key point of interest for swarm robotics. The ordering of this impermanence is what counts.

In the early 1990s, when crowdsourcing and cloud computing were still more theoretical than actual and Wilson and Hölldobler's monumental compendium of ant biology, *The Ants,* was winning the Pulitzer Prize and drawing unprecedented attention to the vicissitudes of formicid life, Marco Dorigo, a pioneer in the development of swarm robotics who now co-directs the artificial intelligence lab at the Free University of Brussels, was finishing his PhD in Italy. Dorigo proposed what proved to be a groundbreaking "ant colony optimization" algorithm for his doctoral thesis, modeled after the way worker ants solve transport problems by leaving trails of pheromones behind as they explore potential paths from the hinterlands to the nest, a method known as stigmergy.

Dorigo credited an experiment by the Belgian biologist Jean-Louis Deneubourg as inspiration for his own work. Deneubourg set up two "bridges" between a food source and a captive colony of Argentine ants, the tiny, ubiquitous world travelers that tend to show up in our kitchen every summer, black threads trailing under the windowsill. Over time, in a random and uneven fashion, the ants traversed the bridges and laid down scent trails, until one bridge gained a stronger chemical signature than the other and became the nest's preferred route. Further experiments showed that the ants could select the shortest route to the food source using the same crowdsourcing mechanism: as time passes, the chemical signature of the shortest route stands out from the rest as more and more foragers mark their passage along it, while the traces of pheromone on the less viable trails fade away with disuse. In essence, the colony has developed a biochemical method for assessing an array of possible outcomes and selecting the best from among them, not by bringing information back to some central decision-maker but by exchanging messages locally amongst the foragers.

Dorigo's key insight was to apply this mechanism to a classic conundrum known as the "Traveling Salesman Problem:" if a salesman has to visit a number of cities, what would be the shortest route to connect them and return to the starting point? To envision Dorigo's response, it might help to imagine a computer screen divided into a grid, with flashing dots to represent the ants, positioned where horizontal and vertical lines cross. These virtual "ants" have been programmed to move along the lines of the graph, choosing and exploring potential routes between destinations, recording the distance they've traveled and the time elapsed as they scurry back and forth between the two points. The data they collect serves as the virtual equivalent of the pheromones ants use to mark their passage, and since the data is time stamped, these markers can "evaporate" as they time out, further accelerating the emergence of an optimal route from the rest. The route these ants discover is not guaranteed to be the best—to verify that would require traveling every possible square on the graph, even those that are obviously not going to be helpful. Instead, Dorigo's model offers what's known as a heuristic—a timely way of locating, with an acceptable degree of confidence, an answer that works well enough and might even be the best possible choice.

Dorigo's ant colony optimization model unlocked one potentially useful mechanism of ant behavior, but if there's a broader underlying principle at work here, one interesting to researchers in ant biology and robotics alike, it's emergence, an almost counterintuitive idea in this age of microcosmic discoveries about the way things work, be they genetic or subatomic. The physicist Paul C.W. Davies describes this phenomenon in *The Re-Emergence of Emergence*:

> Roughly speaking, [emergence] recognizes that in physical systems the whole is often more than the sum of its parts. That is to say, at each level of complexity, new and often surprising qualities emerge that cannot, at least in any straightforward manner, be attributed to known properties of the constituents. In some cases, the emergent quality simply makes no sense when applied to the parts.

Emergence has thus stood as a longstanding caution to the reductionist impulses of the sciences, the belief that biology can be explained by delving into the underlying minutiae of chemistry, which can then be explained by digging into the underlying laws of physics at the subatomic level and beyond. But emergence itself is really two distinct paradigms, one known as "strong emergence," which can trace its history back to the ancient Greeks, and a contemporary "weak" version which is common currency among those who study swarms.

Strong emergence maintains a firm belief in irreducible phenomena—consciousness is the most frequently cited example. We'll explore further the question of consciousness later on this biomimetic trail, when we meet potentially mindful creatures meant to imitate our animal companions. But briefly, the argument goes like this: Nobody can explain consciousness by examining the neurological ball of yarn from which it emerges, and no AI simulation of our neural networks, no matter how intricate and detailed, can represent what it feels like to be human. This is true of our own brains, but even more evident if we extend it to other species, like Nagel's bat. The biomimetic translation of animal to machine can never succeed by literally copying the physical attributes of the original. "If there are phenomena that are strongly emergent with respect to the domain of physics," the philosopher David J. Clayton contends,

> then our conception of nature needs to be expanded to accommodate them. That is, if there are phenomena whose existence is not deducible from the facts about the exact distribution of particles and fields throughout space and time (along with the laws of physics), then this suggests that new fundamental laws of nature are needed to explain these phenomena.

In calling for new fundamental laws, Clayton is not necessarily embracing mysticism so much as the potential for a new set of rules and mechanisms that don't depend on particle physics, but nevertheless you can sense the mystical, almost theological appeal of such a view, the belief that there is something transcendent animating the details of existence, or if not animating, then invigorating them, making them matter. Indeed, the early 20th-century philosopher Max Weber once lamented the loss of enchantment in modern life brought about by the advent of logical and rational explanations for previously wondrous phenomena, a complaint that only amplifies earlier struggles between the spiritual and the rational, as in Emerson's famous opus "Nature," in which he struggles to reconcile "carpentry and chemistry," the stirrings of scientific inquiry and rational explanation, with his desire for a universal spirit permeating, and enchanting, all things. A belief in strong emergence answers the call for "re-enchantment," for the existence of phenomena that remain mysterious because they defy the logic and methods of the prevailing scientific paradigm.

However, what feels "uncomfortably like magic" comes with a whiff of intellectual disdain—any recent consideration of the merits of strong emergence hastens to quash a connection to Vitalism, a school of thought, still wildly popular into the 1920s, which believed the spark of life was a strongly emergent property, an essence circulating among the living that couldn't be reduced to the rigid formulae of chemistry and physics. The discovery of genes as the biochemical lattice of life so thoroughly extinguished this view, however, that Vitalism still serves as a pejorative in certain circles, a shorthand accusation of archaic solipsism.

Yet the discovery of the double helix wasn't the end of emergence—far from it. Dorigo's ant based model exemplifies the return of emergent thinking in an enchantment-free, engineering-friendly and far more limited form. Weak emergence refers to the unexpected connections like those between a single locust and a horde: you wouldn't predict, based upon the simple properties or behaviors of the individuals in a group, the complexity of what you actually observe happening. One minute a host of hungry locusts might be milling around, jostling and nibbling each other in a chaotic mass of mandibles and springy legs, and the next a "rapid transition" occurs and they're marching in regimented fashion toward a crop field. At the outset, there seems to be no logical path between the original state of the constituent parts and the end state of the group. In this sense, the weak and strong versions are alike—both refer to the surprising or unexpected. But in the case of weak emergence, that surprise is temporary and limited. The article of faith here is that what appears to be a frenzied scrum is actually a mechanism proceeding according to its own rules; there's a "natural algorithm" at work, a hidden method to the madness, a measurable sequence of actions and reactions, forays, and adjustments from which the group behavior ultimately emerges. No new fundamental laws of nature are needed.

Wildness, it has always seemed to me, is an emergent property of the strong variety. It runs through the canon of nature writing, vivifying both landscape and creature alike, an epiphany irreducible to the atomic particularities of matter, an "occult relation" between the human explorer and the "real matter" of "rock air fire and wood." We can encounter it with Edward Abbey, posing and then answering the question of why we might venture voluntarily into the inhospitable desert:

> Across this canyon was nothing of any unusual interest that I could see—only the familiar sun-blasted sandstone, a few scrubby clumps of blackbrush and prickly pear, a few acres of nothing where only a lizard could graze, surrounded by a few square miles of more nothingness interesting chiefly to the horned toads.... But there was nothing out there.... Nothing but the silent world. *That's why.*

The premise of much of nature writing, it's central epiphany, involves this kind of decentering of human experience, the "democratization" of the living and nonliving, the sense that we're not the lead actors on the stage, and the stage itself has meaning. The desert exists out there whether we're looking at it or not, and our perception of it, our knowledge of its inner workings, isn't the be-all and end-all of its being. Local rules, in essence, may add local color, but the details uncovered by research into weakly emergent phenomena will never fully encompass the being of objects that are out there. The wild mysteries of strong emergence might thus be taken as a sign of our own human limitations; it's not that we haven't discovered all the complex permutations of the universe's operating system yet; it's that we can't ultimately know them all. There is no universal explanation for all things. In the end, it's beyond us, this silent world, our perception like a stone arrow pointing to the vista that stretches into the distance. The difference between science writing and nature writing is a reflection of the difference between weak and strong emergence.

Of course, Abbey was a defender of wilderness, a place defined by the absence of anthropogenic creation. Phones, drones, wheels, tents...anything we've made, essentially, just gets in the way of getting in touch with what's out there, which doesn't leave much room for biomimetic inspiration. Conventional wilderness is a technological dead end, and as we find ourselves in the midst of the Anthropocene, enmeshed in a climate of our own making, we could legitimately ask whether such a space actually exists. But Abbey's hike itself, out to that lookout point across the canyon, was also a practice of engagement with the nonhuman world. Maybe we need to distinguish here between the practice of the wild and the place known as wilderness. Instead of the hallmarks of conventional wilderness, consider the implications of a list drawn from the collection *New Materialisms*: "tumbleweeds, animal species, the planetary ecosystem, global weather patterns, but also new social movements, health and crime." Our walk transpires in this expansive and "entangled" landscape, where perhaps we can trace the wild as it seeps and flows and emerges even in the most unlikely places.

Karen Barad offers a model for how things work in this landscape that seems particularly applicable to the emergent behavior of swarms. Her "agential realism" challenges the presumption that living and nonliving "agents" exist in and of themselves; instead, Barad argues, it's the relationships between the prickly pear, the sandstone and the curmudgeonly hiker that defines their existence in the moment, that are so deeply formative that they should be called "intra-actions" rather than interactions. As if in answer to Clayton's call for new fundamental laws of physics to explain emergent phenomena, Barad grounds her relationship-based paradigm in a reading of the quantum theory of Niels Bohr, arguing that the primacy of relationships between particles at the atomic level extends more broadly to all states of being—in essence, there are phenomena that cannot be predicted by describing the initial characteristics of individual particles or individual members of the swarm. Instead, the relationships between these entities happen first; even the most solitary figure is defined by the context that surrounds them. The example Barad offers is a slime mold colony, composed of tiny unicellular organisms, whose collective behavior produces the phenomenon that Wilson and Hölldobler described in ants, the superorganism. Barad's point takes the emergent character of the superorganism one step further—the superorganism brings into being the slime mold cells of which it is composed; they don't exist independently before merging into this collective phenomenon. Properties *emerge*, in other words, that cannot be predicted by the initial state of the members of the colony—you can't derive the superorganism phenomenon by dissecting it down to the cellular level. Only as those members begin to interact, in the parlance of weak emergence, or "intra-act," in Barad's version, do things get interesting.

According to the logic of weak emergence, the superorganism's "intra-actions" seem like they might actually be "natural algorithms" that could be represented by an algorithmic model. Barad, however, is wary of representation, of the impulse to make the model itself definitive at the expense of "real matter." The model, she argues, is static rather than dynamic, a dry reduction of what actually happens in the moment. She wants things to exist, not as models or figments of language, but tangibly, as relationships performed in space and time.

Indeed, Barad describes "intra-action" as a performance—the agents perform their rituals and communications in real time, producing the collective phenomenon that is the swarm or the slime mold superorganism. Each time they perform is therefore potentially unique in space, chemistry and time, and thus the algorithmic model is always going to be a heuristic, a working script, good enough but still a static approximation of what is actually a dynamic performance, an impermanent phenomenon.

This emphasis on the emergent properties of performance reminds me of the tension between freedom and representation in Snyder's definition of wildness. Snyder describes the wild as "eluding analysis, beyond categories, self-organizing, self-informing, playful, surprising, impermanent, insubstantial, independent, complete, orderly, unmediated, freely manifesting, self-authenticating, self-willed, complex, quite simple. Both empty and real at the same time." To a degree, that could describe a slime mold or a robotic swarm in its mixture of simplicity and complexity, its self-organizing defiance of categories, but notice the emphasis on the autonomy of the self, the independence, the refusal of mediation. There's something elusive in this definition that the algorithm, or even a line of verse, can't fix in time. "In some cases we might call it sacred," Snyder adds, crossing over into "magical" or theological terrain.

Barad's skepticism of representation extends explicitly to biomimetics. She rejects the impulse to mirror nature, preferring instead a practice that feels more like bioinspiration. "Contemporary practitioners of biomimesis do not claim to be making replicas of nature," she argues. "Rather, they are engaged in practices that use nature as inspiration for new engineering designs." And soon after: "Biomimetics honors Mother Nature as the primo engineer, but it doesn't promise to abide by her methods. It embraces new innovations, new materials, new techniques, new applications. Bringing the new to light is its highest principle." It's not just that we *can't* mirror the subjectivity of a bat, but that we wouldn't *want* to, because we don't want to be forced to abide by Nature's example when we're trying to invent something new.

I'm skeptical of this skepticism, because I suspect there's more going on among practitioners in the field than the disavowal of the mirroring impulse and the devotion to the new. The same project brings together different disciplinary approaches—the biologist and the engineer collaborate in the creation of a model, an extended metaphor for what evolution has devised, but that doesn't mean they have the same objectives. One goal might be to mirror the biotic system to understand the biology better, a process known as "artificial ethology," while another might be to use that mirror as the start for something further afield. Those goals aren't mutually exclusive: biomimetics and bioinspiration might shape different properties of the same creature.

This egalitarian sense of what matters keeps us traipsing along here too. A biomimetic nature walk takes you into territory that is neither conventional wilderness nor a megamall of all things—it's an ecotone of sorts. It's the stringent orderliness of Argentine ants, and the mathematical rigor of Dorigo's algorithms. Maybe the wild lies between them.

~

As my plane banked and descended through a hazy November sky toward the Phoenix airport and the Arizona State University campus, it felt like I was heading into the unknown. Not literally—I could see the shimmering tarmac down there, surrounded by the nothingness of Abbey's desert—but in terms of finding wildness. I had always searched in a familiar direction. Park the car at the trailhead. Hoist the pack. Step into the wild. Or a more paradoxical version: head out into the city, find the ironically wild in the urbane, in the most densely anthropocentric of environments. In either case, push farther and deeper, past the surface, and some sense of it, a glimpse at least, is there.

That's the narrative I knew; it's the conventional narrative arc of nature writing. But weak emergence had thrown me for a loop. What remains after the discovery of hidden conformity to existing laws? Is Barad's performative practice an escape route from this ancient tussle between the mystical and the mechanistic? Are her agents the "free agents" Snyder describes? Where do you find the wild in a mechanistic universe?

I was on my way to visit Stephen Pratt, who began his career teasing out the behavioral mechanisms of ants under the tutelage of Wilson and Hölldobler at Harvard, receiving his undergraduate degree in 1988, just as Dorigo was deep in the development of his ant colony optimization model, biology and computer science coming together in the study of complex systems. Pratt has what seemed to me like an ideal background for this consideration of swarms both biotic and robotic: after his undergraduate focus on ants, he got his PhD studying honeybee behavior at Cornell, then joined the faculty at Arizona State University, where the model of biologist and roboticist collaboration holds sway. Arizona State even offers a master's degree in biomimicry.

Tempe seemed to be enjoying a mild season, although I'd hesitate to call it autumn in that desert climate. The campus was bathed in saturated light, with rows of statuesque palms and live oaks giving it the feel of an oasis surrounded by broad highways that stretch toward the Phoenix airport and the sere rusty furrows of the mountains beyond.

Pratt met me at the door of the science building, and even though I'm several inches over six feet, he bent slightly to shake my hand. He's exceptionally thin, with straight, smoothly parted brown hair, which along with thick glasses, pressed grey trousers, and a tasteful collared shirt, gave him the air of a biological sciences version of Clark Kent.

At first glance, the only clue to the formicultural devotion of this lab space was an abundance of clear polycarbonate paraphernalia. There were bins full of lids and containers, many of them empty, some labeled cryptically with masking tape, others sprinkled with what appeared to be dead ants, like the crumbs left over in a lunch box. It was a space defined by scientific inquiry into the phenomenon of weak emergence, but this tableau of intriguing specialized objects immediately had me wondering about Bogost's practice of engaging with the material world by creating the peculiar lists he calls ontographs.

I could create an ontograph of Pratt's lab; in fact, I already was. I was jotting notes as frantically and surreptitiously as I could and fixing further details in my mind. Lab bench, stool, white ceiling lights, shades over windows, clear boxes full of clear boxes…

As I compiled my list, Pratt picked up an apparently defunct case that looked to have a few dead specimens inside.

"Let's start with these guys," he said.

There was no soil inside, no dirt, no clay or sand to tunnel through or form into little puckered mounds. Most of the inhabitants appeared to have accumulated inside a block of balsa wood with a disc drilled out in the center, forming a hollow space about the size of a golf ball. The disc was notched on one side to connect it to the rest of the container, where a test tube was rolling around with a wad of cotton as a stopper and what looked to be water seeping out from inside. Nearby was a scrap cut from a plastic dish made for weighing powdered chemicals with a smear of something brownish on it, dotted with one or two immobile ants.

"I've had people ask if they were dead," Pratt said, as if reading my thoughts, and gestured for me to look more closely as he held the dish up to the light. I saw one worker twitch to life and amble around for a few steps, sluggishly, like an old dog looking for a new spot to snooze. They weren't dead after all, just loafing.

These were *Temnothorax rugatulus,* which belong to a genus that's widespread across North America. Many species in the genus are known for the relatively diminutive size of their colonies. One eastern *Temnothorax* species Pratt studied extensively before he arrived in Arizona, for example, may squeeze an entire colony into the rounded dome of an acorn. *T. rugatulus* colonies, which typically number about a hundred individuals, are common in the cooler high country, two hours of driving northeast of Tempe, where they nest in crevices between rocks. They're easy to find, and easy to propagate and maintain in the lab, since they don't seem to mind the sterility and lack of privacy. Even the bright light illuminating what should've been the dark cave of their nest didn't faze them. The colony Pratt was holding had been going for at least two years, he said.

Pratt slid one of the nest cases under the microscope. Magnified under the bright light of the scope, the workers glistened, the gaster, or the bulbous rear end on each ant, catching the light like a translucent topaz bead. I could see the head, the jaws, the short antennae, all poised as if about to move, but caught in a state of suspended animation instead.

"You see the larger one—that's the queen," Pratt said, nudging the case slightly with his finger. The queen was darker than her worker sisters, her curves more pronounced. The brood looked like creamy overstuffed bolster pillows scattered around the inside of the disc. Every so often one of the workers gestured with its antennae, picking up on the hint of some message in the air, or shuffled a few steps. None entered via the entrance that was notched into their disc, and none exited to conduct a bit of foraging business outside. They stayed put—if there was a couch, they would have been sprawled on it. It made me think once more about the presumptions we make about what it means to be wild, and about Dorigo's optimization model, and how easy it is to anthropomorphize social insects.

"Not the most charismatic," Pratt admitted. Still, he said, their lack of entertainment value is partly why they're so useful as research subjects. The way myrmecological research often proceeds is by marking individual ants with paint, usually by immobilizing them and then dabbing different colors on each of their segments with a tiny brush, one for the head and thorax, two for the gaster. The ants may have been performing their colony functions all along, but the paint makes it seem like a performance for us, since the effect, when a colony of ants is all brightly painted, is somewhere between a pointillist painting and a miniature carnival parade—Pratt himself once won a science photography award for a snapshot of *Temnothorax* daubed in gaudy colors. Once marked, a video camera records the doings at the nest over a period of time, which can be rewound and replayed as necessary. A colony of a hundred lazy ants is just right, basically, for research: big enough for emergent behavior to occur, small enough to keep track of the individuals in the swarm, sluggish enough that you don't have to rewind as often to see where they're headed.

These qualities have allowed Pratt to study several collective decision-making behaviors that also happen to be useful in swarm robotics. When *Temnothorax albipennis*, a closely related species, move their nest site due to damage or disturbance, for example, they decide where to move collectively, via what amounts to a set of "in case of emergency" instructions. Pratt and a team of researchers studied these emergency rules by first disturbing a nest and then observing the mechanisms of site selection that ensued. They found that once the scouts locate what looks like a promising nest site, they strive to recruit fellow workers to come check it out, first by going on "tandem runs" that lead a potential recruit all the way to the site. At a certain point, however, the ants switch from recruiting via tandem runs to actually picking up their lazy relatives and hauling them to the new nest on their backs. Pratt has put some video of this behavior online—it begins with what looks like a wrestling match, until the smaller ant winds up flipped backward into the air with its head still clamped in the jaws of its sibling, who then trots hastily to the new nest. This switch is triggered by what the researchers termed "quorum sensing:" the ants recognize they've reached a critical mass of workers at the new site who are all "convinced" that it looks good, and they respond to this sense of broad agreement by carting the rest of the colony to the new site as quickly as they can. Ultimately, this consensus building process allows the ants to select the best new site from among several options, even if, as a subsequent study revealed, the colony is initially divided among competing sites with different amenities, like a larger entry crevice or less sunlight exposure.

What impresses me is the degree of immersion in the minutiae of ant life, right down to the painting of individuals, one by one, with a single fine strand of a brush, and then the translation of the observations into a mathematical metaphor. It's easy to see the performance in such approaches, the biologist acting like the set designer, the ants performing their collective behaviors. We might think of this staging and observation as the first act in a biomimetic performance.

~

After returning the *T. rugatulus* to the serenity of the lab bench, we headed to an adjacent space—the myrmecological faculty had arranged their labs contiguously, not unlike the chambers in an ant nest. This one was dominated by the digs of a local species of harvester ant, *Pogonomyrmex californicus*, whose colony occupied two troughs of sand that ran ten feet across the middle of the room. We looked in on tiny "minims" of *Acromyrmex versicolor*, desert leafcutter ants tending their fungi gardens that looked like clots of oatmeal sprinkled with paloverde leaves, and smelled like sourdough. Next door were colonies of *Myrmecocystus mimicus,* desert honeypot ants. Now I could feel metaphorism starting to kick in, the figurative wheels in motion. This felt like a nature walk, all in a single room. Pratt lifted the sheets of light filtering red cellophane so we could peer inside at the "repletes" clinging to the sides of the dish as if it was an underground cavern. Their gasters ballooned with nourishment like tiny golden currants—it would have seemed grotesque if it wasn't such an ingenious counterpoint to the *A. versicolor* grow-your-own strategy. I was reminded of Rolf Müller's collections of anatomical biodiversity, the nose and ear zoo. Müller was doing his own biological research by necessity, but here the tasks were divided by discipline, the mechanical engineering happening elsewhere.

In the back of the room, a graduate student was rolling and rewinding a magnified video of ants crossing a granular patch of sandy soil, taking note as one and then another moved in and out of the screen. Pratt motioned me to an adjacent desktop and pulled up a video clip in which a cluster of slender ants with long, almost spidery legs was tugging insistently at the edges of a yellow plastic chip, which had been rubbed with the scent of overripe fruit. These were desert harvester ants, *Novomessor cockerelli,* which are known for their prowess in hauling cumbersome food items, like ripe figs, that have fallen to the desert floor. This species assembles teams of "porters" to carry the prize back to the nest intact rather than carving it up into smaller pieces for individual transport. They refrain from chopping it up not because it's more efficient, researchers believe, but to get the food underground more quickly, before competitors show up to steal it.

As we watched them wrestle their fig-scented foam chip, some *N. cockerelli* moved to one side of the chip to push; others rotated to the opposite side to pull, while others appeared to show up for a status report, taking stock of things before departing, possibly to recruit new porters to join, or to lay a pheromone trail back to the nest. The load itself moved forward and sideways and around, jerking this way and that way but steadily onward nonetheless, like a king mattress carted awkwardly through a doorway by a team of furniture movers.

"These are the ones that really got the engineers excited," Pratt said.

It was easy to see why, to imagine a scenario in which a team of quadruped robots is asked to do pretty much what these ants do with a fig: pick up a heavy object, like a shipping container or a downed plane, and deliver it somewhere in one piece. Think of it in terms of mattress-moving: if you've ever been on the downstairs end of a mattress, you know it's a nightmare—somebody is either pushing when you want them to stop before they break your back, or they're pulling...in the wrong direction, while your knuckles are grazing the floorboards. You can't see each other, and you have no system in place for signaling each other physically. You just wing it, and probably fling a few curses up the stairs.

When ants show up for some heavy lifting, however, they have a system. Pratt and his robotics colleagues set out to identify it, and then apply it to a similar problem that a group of robots might face. Imagine a scenario in which, as his robotics colleague and lead author Spring Berman described it in the *Proceedings of the IEEE*, multiple robots are

> tasked to manipulate an arbitrarily shaped payload, which is too heavy for a single robot to move, to a target destination without a priori knowledge about the payload or obstacles in the environment. The robots must rely on local sensing and no explicit communication in order for the strategy to be scalable.

First, the research team had to identify the local rules governing the ants' behavior, which meant breaking down the task into deceptively simple steps. The ants have to approach the object. They have to surround the object. And then they have to push and pull the object to move it. Each of these steps involves a mechanism that generates emergent behavior in response. The ants need a recruitment strategy, for example, to gather an adequate number of porters. They need a consensus building strategy to decide where each ant will assemble around the awkward object, some agreeing to push, others to pull. They also need to coordinate, once they get going, to keep the fig from skidding off course like a dropped mattress every time there's a bump in the road.

N. cockerelli are known to have several recruitment and consensus building strategies in place. They first use a variation of the stigmergic pheromone strategy which Deneubourg described, but instead of tracing a chemical path to the nest, they deposit daubs of scent around the perimeter of the fig, which triggers a reaction in their peers, who respond by "suddenly changing course and moving in a zig-zag pattern toward the pheromone source, where they join in attempting to move the prey and releasing more recruitment pheromone." If that doesn't gather enough support, a forager will follow the more conventional procedure of laying a pheromone trail back to the nest and recruiting more porters to join the effort.

Once the ants assemble around the fig, they start experimenting with different angles and positions, until they eventually hit on a configuration that works, and the fig begins to move. The rotating and revolving, slightly meandering path they take is actually a physical expression of emergent properties at work—the ants are constantly in a contingent state, re-assessing and testing and converging on a direction but without the full commitment that a command from above might provide. If they encounter an obstacle, this method allows them to adjust on the fly, rearranging themselves until the fig starts moving sideways around it.

One way to get a handle on what's going on is to look at the force dynamics surrounding the object, as you might if you were accusing your fellow mover of not doing enough with their corner of the mattress, only to have them reply that *they* were actually the one doing all the heavy lifting. Knowing where the ants are applying forces, the direction of the forces, and how much force is being applied can help reveal the mechanism that ultimately gets the fig in motion.

To identify it, the team devised an experiment that took place outside the entrance of a nest in the middle of the desert, where they placed a block of wood covered in sheets of paper to provide a smooth surface. On it, they placed a circular foam chip, to which several carefully calibrated "springs" made of clear elastic material had been attached. At the end of each spring was a tiny rectangular handle, which they rubbed with fig paste. These handles allowed the ants a convenient and attractive grasping point, while also limiting the range of the possible positions around the chip.

Video cameras recorded the moment when a forager detected the scent of fig and the whole retrieval process unfolded, with ants tugging on the springs in increments that could be measured and turned into a data set by integrated software that tracks the movement of visual targets in the frame, which in this case was a central black dot along with the rectangles attached to each of the springs. The accompanying photos show one or two of the deluded porters taking up their position at the spring tabs, their whiskery rear legs braced with the effort. It's an odd juxtaposition of biotic and geometric, made more so by the fact that this encounter was happening in the middle of a Sonoran arroyo, like a film set on which improv actors do their best with implausible props. *Pretend this is a fig...*

The results indicated that the carrying process advances through two phases, with the ants starting out slowly as they come to grips with the logistics, and then proceeding down a faster and straighter path once they all get on the same page. The researchers proposed two models for how this cooperation transpired, the first of which captured the squeezing and stretching of the springs as the ants figured out how to work together, while the other mapped the movement patterns of different ants as they oriented themselves around the load. Provisionally, at least, the first step toward explaining a weakly emergent phenomenon was complete.

The next step was simulation. To test the validity of these models, the team created a virtual environment in which the physical elements, the load, the springs, and the ants, could all be represented. If the team observed actual ants trying to rotate the object so they could walk backwards, for instance, then that action would need to be possible within this virtual reality, reproduced as a range of potential angles specified for each movement of the simulated "ant." Time is also an important part of these equations, since collective behavior emerges from random trial-and-error as a function of time passing, so the opportunities for the "ant" to pursue different movements must be specified as a sequence of possibilities over time.

In an accompanying screenshot of the simulation in progress, the assembled "ants" look like simplified legless pegs with a black dot for a head, oriented at different angles and distances around linear springs that jut out like spokes from the circular representation of the load. If the real ants looked like actors on a film set, these guys looked like characters in a very simple cartoon.

The important point, however, from a research perspective, was that the team was able to replicate the results of their experiments in the simulation—the virtual ants behaved like real ants, within the limited expectations set out by the models. The team had taken a significant step toward getting robots to behave like desert harvester ants, a useful development for engineer and biologist alike.

There was one additional item of significance, at least from my perspective. These were the one species that wouldn't perform in captivity. In their desert habitat, they can be fooled into transporting a piece of fig-scented plastic as if it is a piece of fallen fruit. But nobody has yet figured out how to make them do so in a laboratory setting, even if they are given a large piece of dried fig. If you want to study the fig-bearing behavior of *N. cockerelli,* you have to find them on their turf, and accommodate their rules. They yield information, but they also remain resolutely outside the box, absorbed in their own version of the world, their own ordering of impermanence.

~

Virtual simulation is ubiquitous and relatively cheap; real world testing on the other hand has always been a time-consuming and expensive proposition, which is partly why many research projects end on a computer screen. Of course, there's potentially a world of difference between a virtual simulation and an embodied one. In translating the behavior of actual ants into a "natural algorithm," and then translating that algorithm into the "novel algorithm" that works in a virtual landscape, the possibility arises that anything you overlooked, any hard to observe or categorize ant behavior that didn't make it into the data, anything you simplified to resolve some technical difficulty, all might be reinforced by the simulation. You might, for example, avoid the messy business of collisions by having virtual ants simply move right through each other and carry on as if nothing happened. Or maybe they just "teleport" to a new location. In simulation, that works just fine. It's tempting to tune out the "noise" of inconsistency until the model appears to work on the screen, and everything appears to comply with the parameters you've included in the model. But does the algorithm actually mimic what the ants were doing as they wrangled their fig through the noisy, less-than-ideal physical world? Would the model you've devised for robotic creatures work in real world conditions? Surprises may await when the virtual meets the physical—when the software that works so well in simulation is actually embodied in the mechanical equivalent of legs and antennae.

That's where the Robotarium comes in. Magnus Egerstedt, who directs the Institute for Robotics and Intelligent Machines at Georgia Tech, created this specially designed amphitheater as a "24-hour ecosystem" where a swarm of a hundred small robots perch on chargers, waiting for instructions from researchers who might be halfway around the globe. It could be an ant biologist, or a roboticist studying heterogeneous swarms of terrestrial bots and aerial drones, or even a nature loving neophyte like me—the idea is to democratize access to the tools of biomimetic research and bring these experiments one step closer to application in the real world.

I first met Egersedt in his office one afternoon in an Atlanta neighborhood whose construction noise and scaffolding spoke to its rapid transformation into a gleaming mix of high-tech incubation and upscale dwellings. His team of graduate student researchers was also feverishly at work, putting the finishing touches on the new and improved space while still performing experiments with the prototype. The mix of fatigue and frenetic activity seemed to extend to Egerstedt himself, who after inviting me to take a seat on a bright red sofa, immediately turned to extracting espresso from a fiery red capsule machine. The couch faced a giant whiteboard, on which intricate mathematical constellations headed toward the ceiling, circular graphs and algorithmic abbreviations punctuated here and there by snippets of the English language. In another corner, a Pinocchio-like puppet leaned against the wall in a plexiglass case, a memento of a prior study of the robotic choreography of marionettes, but also a nod to the notion of animating the inanimate, of life quickening in the limbs of something that is ostensibly just a replica, strung together with lifeless materials.

Egerstedt brought the coffee in two tiny cups. He was dressed in loose jeans and an untucked lavender oxford, his buzz cut brown hair and close-cropped goatee overshadowed by a pair of thick framed glasses whose round lenses seemed to magnify the gently ironic expression in his eyes. He's originally from Stockholm, and he speaks with just the faintest cadence of a Swedish accent, often punctuating his points with an emphatic, eyebrows raised "Dude!" which causes ripples to appear in his forehead.

He'd begun his career in a roundabout fashion, as a philosopher studying the abstract riddles of individual consciousness, an area known as the "theory of mind." "I grew up asking questions of what it means to think," he said. "But I was also fascinated by these mesmerizingly beautiful shapes, like the beauty of fish schools. How do fish form these shapes? I'm a geometric guy." He turned to applied philosophy, to building systems to answer empirical questions about how these shapes in schools of fish or flocks of geese develop, and somewhat to his surprise, began to publish his findings in robotics journals. Emergence, he said, is the driving force behind his work:

> Nature is absolutely filled with examples of such systems, these cases where there's too much going on for the normal channels to work. No central brain can handle all the work. We're drowning in complexity--we have to go to local rules.

In describing his vision for a space built around physical movement, Egerstedt mentioned the influence of the "bottom up" ideas of Rodney Brooks's position paper, "Elephants Don't Play Chess." Brooks keeps reappearing on this biomimetic journey; his paper, like Nagel's essay on bats, reads like the field notes from an early explorer, a seminal work for those advocating close attention to the behavior patterns of creatures that evolved to deal with real world problems.

Individually, the Robotarium's swarmbots seem like descendants of Squirt, the early robot built in Brooks's lab. It was a simple creature, a tiny cube with motorized wheels and not a lot of brainpower, but it was nevertheless a bundle of perception and action, grounded concretely in the world. What's different about the swarmbots, of course, is evident in the plural pronoun—we're talking now about complex behaviors arising collectively, as if Squirt had brought a whole gang of friends along. Like Squirt, they interact with the world, but that world is now teeming with other robots. These robotic swarms can develop as assemblages of drones, wheeled robots, and humans in the manner of different species interacting to form ecological relationships.

I used the term biomimetic to describe the local rules that govern those interactions, like the rule that makes Squirt run when it senses light, but Egerstedt subtly corrected me, using the term bioinspired instead. He might begin with a behavioral mechanism drawn from the riches of the biotic realm, he said, but he wouldn't necessarily worry about whether the robot behavior ultimately matched the ant behavior. The biology, for him, is a point of departure, not a goal. Rather, his focus is on the robotics—could a novel algorithm, inspired by the emergent behavior of insects, enable robots to solve problems? His research might begin by finding models for local rules in the pages of biological journals, the kind where Stephen Pratt's research might appear. From there, he might develop a software-based model that approximates these local rules, then observe what happens as the robotic agents interact, under physical conditions that might not mimic the natural world. He might, for example, have a pod of robotic "dolphins" compete against a pride of robotic "lions" to see which could catch a "prey item" first, using their collective behavior strategies.

"I am a shameless user of biology," he said, jokingly. "I pick what I need."

For a biologist like Pratt, on the other hand, the point of developing an embodied model is to discover something new about the biology in question. Pratt had told me that biomimetics is a useful term in his field—if you're trying to understand behavior patterns in a particular ant species, then you're probably committed to representing the biological conditions that produce it as specifically and faithfully as possible. Ideally, the technology is a mirror in which certain details appear more clearly in the reflection. Just as constructing a robotic lobster, for example, has allowed researchers to study how actual lobsters navigate the ocean floor, the same would be true for robotic ants modeled after the desert harvester ant porters hauling a fig around. Even within the biological sciences, however, the value of mimesis depends on the questions you're asking. If the goal of your research is to understand an evolutionary principle, something that transcends time and species categories and might even link lobster and ant, then you might chafe against the constraints of mimetics and be more amenable to the term inspiration.

In practice, however, biomimetic and bioinspired visions coexist, even within the same research project. It's like two paths that start together, sharing the same route, until at a certain point they diverge, the engineering path heading off into the inspirational distance, while the biological path doubles back to the mimetic start. For example, a team of ant biologists led by one of Pratt's former PhD students, Takao Sasaki, as well as Clint Penick, then a postdoc in ant biology at NC State, developed a model to represent a distinctive behavior in *Harpegnathos saltator,* a species which hails from the tropical forests of Southeast Asia and is commonly known as the Indian jumping ant.

One of the existential crises ant colonies inevitably face is the loss of their founding queen—there has to be some kind of mechanism for replacing her, or the whole colony will disintegrate. *H. saltator* responds to this state of affairs with ritualized sparring between what are known as gamergates, workers whose reproductive switch hasn't been permanently disabled. These wannabe queens, which are relatively numerous in this species, have the capacity to mate with their male relatives and produce offspring that can keep the nest going. But the right to assume this exalted role is limited to a few; they have to establish their reproductive dominance through what amounts to a duel.

On a video posted by Penick, a pair of gaudily painted *H. saltator* gamergates square off, not with the lethal weapons they carry in their gasters, or even with the pair of scimitars that adorn their jaws, but with their harmless antennae, slashing and parrying with these appendages as if they were the whip-like foils used in fencing. Once a small cohort of winners emerges from this tournament of ritualized combat and begins to lay eggs, the rest of the colony rallies around them by "policing" potential rivals—to stop them from developing reproductively, they pin the offender down with their jaws and hold them there until their bodies reverse course and revert to regular old worker status once more.

This model based on the gamergate duel could have been tested exclusively with virtual ants in a virtual environment. Instead, the researchers took the additional step of adapting the model to work with the software that animates the Robotarium's swarm. They played around with the gamergate duel scenarios in a virtual environment, using the customizable ant modules the Robotarium provides. This customization has obvious limits—the team had to simulate a confrontation between gamergates with bots that possess neither antennae nor jaws, so there would be no pinching or flagellating here to determine a winner. A set of probabilities had to be employed instead, essentially a weighted flipping of a coin to determine the victor in any pairing. The "fertility status" of each robot would rise or fall according to the outcome of each coin flip, so that by the end of the tournament most of the lower status robots would quit the field and a cohort of high fertility victors would emerge.

Egerstedt arranged for me to see this robotic version of the gamergate duel in action, just after the new Robotarium space opened up. It was located in an otherwise unremarkable classroom building, but when I opened the door it felt as if I was peeling away a layer of old newspaper to glimpse a new world: here was this gleaming, almost shockingly white space. At the center of the room sat a low rectangular platform made of an almost seamless white polymer of the sort used for high-end kitchen countertops, which curved over the edges almost to the floor. Bigger than a king-sized bed, it looked like a gigantic air hockey table. Natural light was pouring in through a row of big windows along one wall, adding to the daylight spectrum light emanating from recessed fixtures in the ceiling—it felt like a conservatory, a space meant for nurturing orchids and orange trees, except for the multiple motion capture cameras peering down from the ceiling.

The robots waiting for action were meant to be individually simple, each one essentially a rolling circuit board wrapped in a ring of chartreuse plastic that concealed the electronics. On one side was their inductive charging plate, a vertical flesh-colored disc that was easy to misperceive as a face, and on the other was an LED bulb, which could change shades from red to green. They looked identical at first glance, but in actuality their distinguishing feature was a constellation of four plastic pegs, each tipped in a tiny foam ball, arranged inside the green loop. They looked like antennae, or tarsi perhaps, the tapered bit at the end of a cockroach leg that quivers when you find one upside down on the kitchen floor, but they weren't actively engaged in sensing or moving. On each robot, if you looked closely, you could see these projections were configured differently. Their purpose was to reflect the infrared light shining down from above with a distinctive signature, identifying each individual robot for the special cameras monitoring their movements in real time. There were 26 in total, and most were perched in pairs with their charging plate pressed against a blue or yellow induction "target" on the wall where they could retreat from the center of the arena and recharge.

I'd encountered the living version of this species of ant, *H. saltator*, on my tour of the Tempe labs with Stephen Pratt. This, the most charismatic of the bunch, he'd saved for last. At first glance, they hadn't looked much like ants. They were long and strangely slender, as thin as the honeypot ants were round, as if someone had taken ahold of either end of their bodies and stretched them to the snapping point. The thorax in particular looked out of proportion with my own image of what an ant should look like—were I to tie a trout fly based on a generic ant, the head and gaster would be large, but the thorax would just be a dot, a joint to connect the two important bits of the silhouette. The thorax on *H. saltator*, however, was stretched even longer than the other two segments; it looked like a chocolate sprinkle, or the overextended lead of a mechanical pencil.

Pratt had conferred briefly with a nearby researcher and got the thumbs up—it was feeding time. "You'll see how they're closer to wasps," he'd said, flinging a cupped handful of small crickets the color of plucked chicken into what amounted to the gladiator pit.

We'd watched as word got out. A couple sentinels were standing by when the crickets fell from the sky, and they reacted to the sudden appearance of their prey by hustling back to the nest to gather more bodies. The rest of the nest—there appeared to be less than 40 workers in total—didn't pour out like a swarm of angry hornets. Instead, their attack happened slowly, almost by individual persuasion, more and more hunters coming through the tunnel to see what the fuss was about, the dark beads of their unusually prominent eyes picking up the twitch of a cricket leg or antennae, summoning a collective response.

The crickets began to jump. The ants did too, predator and prey pinging off the lid of the case with a soft popcorn popping sound that took me back to keeping grasshoppers in coffee cans as a kid. I'd never seen an ant leap into the air like that.

"You see what it's doing with the gaster?" Pratt had inquired, pointing to an ant that had seized a cricket with its jaws and was now doubled over, extending the tip of its tail to jab into the underside of the cricket's abdomen. The cricket struggled, but it was clamped in the death grip of the ant's specialized jaws, which were like the sharpened and serrated prongs of a fork curved slightly upward. The ant's antennae were pinned back over its eyes so they didn't get bitten off—the mandibles of a cricket, to an ant, are nothing to laugh at.

Over the course of a few minutes, the frantic jumping had all but ceased as the paralytic venom took hold and the hunters began coaxing the first body, stiff and upside down but still alive, back through the tunnel. They'd dropped it just inside, like a mouse left on the doorstep by a cat demonstrating its prowess.

Chartreuse plastic didn't quite capture the ferocity of the creatures I'd seen, but I had to remind myself: this wasn't about convincing me that what I was seeing were *H. saltator* gamergates—this was about representing how ants perceive each other, and what they do in response. It was meant as a performance of what it is to be dueling swarms of Indian jumping ants, which in itself is already a ritualized performance of a battle. We were about to observe a robotic performance of a formicid performance.

One of the Robotarium's postdocs, Sean Wilson, talked me through some of the challenges they'd encountered so far—they'd had experiments come in from every continent except Antarctica, at all hours of the day and night, so the space was often buzzing with drones and terrestrial bots. "Everything's perfect in simulation," he said. "If you say move this fast, it does it instantaneously and you're perfect. Your sensing is also perfect, so that you know where you are at all times in the simulation." In reality, the robots sometimes behaved in ways the model designers never anticipated—they zoomed and zagged at speeds higher than their capacity to make a turn. Netting curtains had been deployed on occasion to keep drones from veering off toward the walls.

Wilson spent some time with the code on his laptop, hit execute and then turned expectantly to the table. The gamergates began to roll, soundlessly, in tight linear formation, out to the center of the table from the charging stations on either side. The two groups merged, but while real ants tend to collide frequently, here there were no collisions; each bot maintained a rigidly defined personal space, like a well-rehearsed marching band.

All at once, they all began to rotate in place, displaying their flesh-toned charging plates in a synchronized move. Then they split up and began to spin around in seemingly chaotic explorations, as if tracing scent patterns on the ground, while from the ceiling the cameras picked up their foam-tipped configurations and transmitted the bots' positions to each other in real time. It put me in mind once more of Snyder's definition of the wild. Yes these robots were "self-organizing" and "self-informing" in the manner of ants. They were to a degree "beyond categories," "complex" yet "quite simple," and the outcome of their performance was "surprising" to a degree. But what kind of freedom, that key feature of wildness, did these robots possess? What's appealing about swarms, from a robotics perspective, is the complex, almost paradoxical nature of their autonomy. They're free from central authority but faithful to layers of control built into collective behavior. That's natural in eusocial insects. But is it wild?

Eventually, some of these robotic gamergates began to face each other in pairs, which Wilson observed with an indulgent, and slightly apprehensive, smile. In actual *H. saltator* colonies, the dueling usually takes weeks to complete, but the team had set a virtual boundary around the center so the roaming gamergates could find each other, speeding up the process. The "duel" didn't last long—it took only seconds for a pair to sense their opponent's relative strength and calculate an outcome before moving on to the next, and it was hard for an untrained eye—mine—to pick out the pairs in the crowd. Wilson pointed to their LEDs—some were turning from bright green to a more muted olive, on their way to red, which would signal their withdrawal from the competition. One by one, the red lights moved to the periphery of the swarm, leaving a small cohort of high-status queen robots in the center, flaunting their bright green LEDs.

I thought about the *H. saltator* ants I saw on the prowl in Tempe, their prominent eyes and spear-like jaws. I thought about the video I'd seen of a gamergate duel, the rhythmic, almost stylized performance that took place within a crush of bodies, more like a rugby scrum than a cotillion. These bots were not ants. This was clearly not a mirror. And yet, what was unfolding was a kind of physical translation of weak emergence from the biotic to the machine. There was still an evocation of jumping ants here, not in the literal physical representation but in the conceptual contours of the performance, in the way you might say that a performance of *Swan Lake* is inspired by the movements of swans, "both empty and real at the same time." These are not swans. And yet, they are swans.

The wildness we've been tracking lies in the space between them.

~

Nine months passed after the discovery of the beehive in my colleague's wall. The last frost passed; the peach trees were in bloom.

Then her daughter heard a buzzing in the hush of early morning. It sounded like a mosquito, but there was nothing in the room. Instead, it sounded like it was coming from the wall.

Her mom put her ear to the wall, and sure enough, she could hear it too, that telltale hum.

Outside, they found a hole in the masonry between the bricks, with honeybees congregating around it. As if, having reconstituted their collective identity in the face of catastrophe, these free agents had retained some collective memory of home, some residue of their occupancy. As if they never left.

The humans called the beekeepers.

"Do you hear buzzing?"

Yes.

"Do you hear the sound of paper crumpling, or something chewing paper?"

No, thank goodness. No.

You have to get rid of every scrap of the nest, the beekeepers said, or they'll be back. If you block the exit, they'll chew their way out. That's the paper crumpling sound.

Try almond oil. Dab it around the entry. If they haven't invested too deeply, they may take the hint.

Almond oil, squeezed from the wilder version of the seeds we eat. An aromatic flavoring, bitter almond, but also rich in cyanide if concentrated enough. Not poison, exactly. A performance of it.

As if, this intra-action between homeowner and honeybee suggested. As if we could send a signal across the boundaries between our species, and you could parse our meaning.

4

INTERPRETIVE STATION

Social robots and other trailside companions

It's a blustery Sunday morning, and I'm out walking the dog. That probably sounds pretty mundane; after all, there are over 50 million dog owners in the United States alone, and every morning you'll find a healthy number of them on our South Carolina street, trundling along at the end of a leash while their canine counterparts immerse themselves in the neighborhood gossip.

But I'm not your average dog walker, and Freedom, although he looks the part of a typical black lab, is not your average dog. While I've hiked many miles of trail in the company of other people's dogs, since childhood I've never actually "owned" a dog myself, so this whole business of taking the dog out to do his business is new to me. And Freedom is no mere pet: he's a service dog in training, and he spends his weekdays in prison, where the lights come on at four in the morning and his trainers lead regimented lives. He doesn't get out for neighborhood walks much, but as his name implies, he's already a source of inspiration for the inmates who work with him, and with luck he'll go on to make a huge difference in the life of a human client, with responsibilities far beyond those of a conventional canine companion.

The transformation from wriggly puppy to service dog requires an enormous investment of time and resources. PAALS (Palmetto Animal Assisted Living Services), the organization that's been training my family to care for dogs as weekend "pupsitters," spends two years on intensive training before pairing the animal with a client. Service dogs need to learn how to focus on the needs of their human companion; they must understand an assortment of verbal cues that might mean turning off a light or pulling laundry out of the dryer, and they have to resist the urge to investigate all kinds of interesting and novel situations. A well-trained service dog is an exemplar of interspecies synchronicity, a Zen master of detachment who won't vacuum up a kibble treat if it falls on the ground at his feet, since it might one day be a pill dropped by a human companion. A dog like Freedom learns from the early days of puppyhood not just to make eye contact but to hold it, to gaze resolutely into the eyes of his companion and decipher the nuances of human expression. Should he successfully complete his training, Freedom will be covered with a life insurance policy of 40,000 dollars.

When I look at Freedom, I see a potential blueprint of sorts, a descendant of wild canines sculpted by the forces of domestication and coevolution into a model for creatures of a different stripe. He's brimming with instinct and emotion and canine intelligence, but he's an anthropocentric animal too, clearly molded by the needs and desires of our species intersecting with those of his own. Somewhere in that mix of free spiritedness and task-oriented accountability lies the sweet spot known as companionship, and that's ultimately what some roboticists are aiming for with software and hardware: a robotic version of Freedom, built to trot along at my side.

That's a pretty hefty intellectual burden for an early morning walk, and Freedom, who at a year old is already halfway through his training, clearly has other ideas about what should be on our agenda. Like running full tilt up the street to a marking spot, then stopping abruptly for a sniff and a splash. The leash is the physical sign of our entanglement; indeed, I'm constantly untangling his limbs and unwrapping my own. Freedom is learning about me, and I am in training to learn the ways of a service dog. I'm trying to learn who Freedom is, what he's like, how he sees the world—or as best I can surmise. He and I have only known each other for 24 hours, and although I've been learning the verbal cues that are supposed to form the basis for our interactions, I still tend to blurt out the wrong word with embarrassing frequency, which usually earns me a quizzical "Are you kidding me?" look from my canine companion. I'm only supposed to say *good dog* three times max, but often my thoughts emerge in an anxious, pleading torrent: *good dog, good dog, good dog…no, no, whoa, Freedom, Freedom, FREEDOM!*

To help me through, I've got treats. Lots of kibble treats in the nylon pouch on my belt. I've been trained to cup one in my backward palm and hold it near my pocket. That's the secret to our partnership: kibble, my little pellets of comfort. I've got a reserve stash in either pocket, even though his food is strictly rationed. I don't want to know what might happen if I were to run out.

The two of us have practiced walking "loose leash" around a parking lot at the PAALs headquarters with ten other service dogs and new pupsitter recruits, but this is the first time we've set foot or paw together beyond the middle of my driveway.

What's out there? Off leash mongrels? Texting and swerving drivers? The hidden cat poo I was warned some dogs might like to eat and then eject out of either end when they got back inside the house? Neither of us knows what we'll encounter, but Freedom is eager to find out. In spite of his training, he's like a bottle rocket about to go off.

No way this could be a robot, I think, gripping the leash with both hands.

But then Freedom stops suddenly and rotates his head, as if he's picking up some faint signal that's inaudible to me. Something about the angle, and his frozen expression, seems almost mechanical. He's processing something; the audial and olfactory sensors are transmitting data.

Will I ever look at a dog in the same way again?

~

So far on this journey into robot country, my feelings have been kind of beside the point. HyQ was born to run; it never begged for my attention, or acted as if it liked my company, or even noticed my existence at all. The same could be said for the batbot sending and receiving its chirps, or the swarmbots performing their gamergate rituals. That's not a knock against their intelligence or autonomy or sensory capacities, but these machines had work to do, and they weren't designed to give a hoot about how I felt.

As we venture deeper into this technological ecosystem, however, a different order of robotic creature awaits us. They're designed to seem as if they crave our company, and to at least gesture toward the notion that they might actually do so. Their central purpose is to push our buttons, to foster the sense of an emotional bond—in us, at least. They're meant to make us feel what we feel when we spend time in the company of an animal like Freedom. An encounter with this kind of creature feels like a different order of magnitude to me, because building a model of "human-robot interaction," or HRI, has to account for us. Freedom and I are both clearly part of the model, this engagement between two different mammalian subjectivities, one human primate, the other canine.

Putting humans into the mix means we're headed for a biomimetic landscape feature which in 1970 the Japanese robotics researcher Masahiro Mori famously called the "Uncanny Valley," where our perceptions of the artificial and the natural form the walls of a canyon on either side of a graph, with the vale of our discomfort down below. Mori's hypothesis was based on observations of the way people responded to different images, some clearly human, others obviously robotic, and then a third category, in which the figures combined elements of the two. With a clearly robotic figure, the observers felt fine; they knew where they stood. That was one side of the graph. The same held true when they observed what was obviously another human being. No problems there. In fact, the line that plotted their approval kept rising as the robots became more and more lifelike; the positive vibes continued to mount until they reached what amounted to a steep cliff of disapproval. Then, after an interval of ennui, the line of positive feelings rose steeply again, forming the other side of the "valley" as the robots performed so convincingly that the observer no longer was able to tell human and robot apart. What made the participants feel ill at ease, then, were the figures that made them do a double take, that appeared human at first glance but which a closer look revealed to be machines.

The uncanny valley appeals to me as a symbol for this biomimetic nature walk, I suppose, in part because it makes a topographic metaphor out of our emotional response to machines. It's a fertile landscape, teeming with biomimetic inspiration, and yet it's also anxiety-provoking. We haven't been here before as a species; we didn't evolve here. While we can find analogues in nature that are elegant solutions to engineering problems, that doesn't guarantee we'll like the result. A journey through the uncanny valley foregrounds a question that has been with us all along: how do these machines make us *feel?*

There's a connector trail between sensing and action where the processing of data takes place, through the mysterious territory of mind and selfhood that stretches past the point where we can't distinguish between animal and machine. It forces us to confront existential questions about the nature of being, not just in ourselves but in machines whose systems are modeled after our ways of thinking, feeling, and relating, the ethereal qualities we associate with selfhood. We are pursuing the artificial ethology of ourselves.

One aspect of this mysterious terrain has already been the focus of considerable attention: the modeling of "intelligence," of cognitive mechanisms that an individual human brain might use to analyze data with logic and reason. These efforts have even been organized around a biomimetic trope, the "artificial neural network," an analogy that brings to mind Otto Schmitt's original use of the term "biomimetic" to describe his research into the electrical activity of a nerve cell. In a biotic system, electrical or chemical signals travel through the network across intersections or nodes that function like switches, diverting them along one pathway or another. This system is organized spatially; a functional MRI might observe which areas of the brain "light up" in response to some kind of sensory stimulus, "mapping" this cognitive geography.

While there is ongoing research into replicating the material structures of individual nerve cells, the term "artificial neural network" more frequently refers to a system modeled after the organizational patterns of rules, hierarchies, and decision trees that enable neurons in the brain to work together, an approach to modeling intelligence known as connectionism. An artificial neural network connects layers of "neurons" in a configuration whose pathways are governed by mathematical rules that resemble, at least conceptually, the connections in a human brain. A network of this sort might address a question about a complex data set by breaking it down into a series of smaller problems, each of which could be handled by a different part of the network, or it might work by amplifying or weakening signals until they reach a threshold, the point where a response takes place, strategies that biological nervous systems like our own are thought to employ.

These models of connectivity also allow for a representation of learning. "Deep learning" programs "train" artificial neural networks to recognize patterns in large collections of data by allowing incremental adjustments to the model as it responds to successive waves of information, a process known as "mapping." An artificial neural network "learns" by first going through a training phase in which trial-and-error and reinforcement figure prominently. The engineers "show" the network a series of examples from a very large data set, and the network responds by sending this information through layers of "neurons," each of which must make a choice between weighted options until the process reaches a final decision point and provides an answer. That output can then be checked against the right answer, which the engineers have already verified. If the model's answer matches the right answer, it confirms the validity of the decision-making structure; wrong answers require an adjustment of the weighted assumptions.

Matching the machine response to the correct answer happens again and again as the program runs, and each time the training program adjusts the model slightly to reflect the previous results, until over time the machine "learns" to produce the right answer with greater and greater frequency. Once the training stage is complete, the model can be released into the wild: rather than testing against a known set of answers, the now finely tuned algorithm can make inferences and predictions in response to live data where the right answer isn't a foregone conclusion. These are incremental steps toward a "data-driven" intelligence that can already outperform us on some analytical tasks, even if the analogy between the functioning of a biotic brain and the way these systems work is more loose inspiration than faithful reproduction.

This process is analogous to the way PAALS trains a dog like Freedom in some ways. They teach the dog through repetition and consistency, by performing the same task until it becomes second nature. Certain responses earn praise or other rewards; others do not. Freedom learns the patterns and responds accordingly. At a certain point, he's taken out of his comfort zone—PAALS may take the dogs to a big box store or send them home with a wannabe trainer like me, so the patterns they've internalized are tested against a new set of sensory data. The trainers use these responses to further refine the dog's training.

To a degree, recent achievements in deep learning validate a neurobiological model of how our brains function, or at least represent some aspects of the way biotic problem solvers and decision-makers like us operate in the world. But the question remains: how far can these steps toward an artificial intelligence take us toward a model of a full-fledged being, let alone a companionable pair like Freedom and me? Freedom seems to do a lot more than ponder; he's not permanently lost in thought and neither am I. We're creatures not just of intellect but of affect, aware of each other and ourselves, even if we are different species. Some experts in AI suggest the success with deep learning has reached its limit—even if intellectual prowess is part of the biomimetic puzzle, connectionism doesn't seem to be the model for the "hard problem" of mammalian being.

~

In the nature writing canon, few moments are more transcendent, more resonant with meaning, than an eye-to-eye encounter with another species. More than spotting a warbler as it flits through the boughs or a bear crashing away through the understory, eye contact is a mutual experience in these narratives, inspiring epiphanies of both otherness and communion. There's Annie Dillard, startled into a metaphysical reverie by a weasel's gaze as she sits by a pond at the end of a walk. "I tell you I've been in that weasel's brain for sixty seconds, and he was in mine," she writes, which leads in turn to rediscovering the instinctual possibilities of human animality, of "living like a weasel." Further down the literary trail there's Aldo Leopold, transfixed by the gaze of a wolf he's just shot. "We reached the old wolf in time to watch a fierce green fire dying in her eyes," Leopold wrote of this encounter. "I realized then, and have known ever since, that there was something new to me in those eyes—something known only to her and to the mountain." We could join John Muir as he peers into the eyes of quail in the High Sierras: "At last one of them caught my eye, gazed in silent wonder for a moment, then uttered a peculiar cry, which was followed by a lot of hurried muttered notes that sounded like speech." Or Edward Abbey crawling along at eye level to gaze into the eyes of amorous gopher snakes as they perform their mating rituals. He's struck by wonder and fear when the snakes come straight toward his face, "their intense wild yellow eyes staring directly into my eyes." A liminal space opens between snake and human as the normal categories dividing the species are momentarily destabilized; what fills it isn't so much projection as transference, a recognition of the self in the other. Maybe domestication mutes or moderates this dynamic; in place of wildness we have a deep sense of intersubjective familiarity, a sense that to some degree our pets are already us, and we are already them. And yet, while a dog is not a wolf, it isn't human either. Domestication adds another layer of complexity to what happens between us, albeit with less drama.

To develop a full biomimetic representation of our relationship with dogs, one approach would be to create a model for the intellectual and emotional life of a dog, and a similar model for a human being, as well as a series of models to describe the mechanisms of interaction between them. Eye contact might be one dimension of this engagement. In fact, eye contact between humans has been divided into different interactive dynamics, including the task-oriented "joint attention" of two people staring at something together, the "mutual gaze" in which two people look into each other's eyes, and "gaze aversion," the avoidance of eye contact, as well as all the social cues that arise by shifting between these modes. Recently, facial recognition software has been adapted to allow a robot to read what our eyes are doing and respond with eye movements of its own, as was evocatively demonstrated by the blue-eyed humanoid robot SEER at an interactive installation in Germany. This robot offered two modes, one of which simulated the mutual gaze, while the other imitated the facial expressions in whoever was engaging with it through movements of the head and eyebrows, as if it was powered by mirror neurons. Even in video form, this animatronic semblance of mutuality was uncanny. But the performance of actions we associate with feeling and being doesn't necessarily address the mystery of actual feeling, especially as expressed in another species.

We could simplify the challenge of interspecies engagement by limiting our inquiries to our own reactions. How, we might ask, do humans respond to the physical appearance and behavior patterns of another creature, regardless of what's actually going on inside it? Or to put it another way, how can a creature's performance of companionship generate an emotional response in us? It wouldn't matter if this performance was actually just a set of animatronic movements or the results of eye tracking software—everything would be in the eye of the human beholder, and the question would be how the illusion made us feel.

We might even argue that this is as far we'll ever be able to go. I'm no dog whisperer, but even if I was, there's plenty of skepticism about whether we humans can ever know what it means to be a dog or any other animal companion who seems to look at us adoringly. Take the famed French theorist Jacques Derrida's cat, for example. Derrida tells a story of toweling off after the usual human bathing ritual, only to find himself naked under the gaze of his household feline. What does the cat see, he wonders—does it register his nakedness? Should he feel ashamed? He knows this cohabitating pet to a degree, he realizes, but that knowledge can only take him so far. Between two individuals of two different species lies an "abyss," a chasm of unfamiliarity at the edge of domestication. We can infer a great deal, this line of thinking goes, by learning about the biology and behavior of felines through observation, by training individual animals to behave in ways that work with us, and even by shaping the physical and behavioral tendencies of these pets through selective breeding. But as with Nagel's bat, there's a limit to what we can know of subjective feline existence, and hence any model of interspecies engagement would necessarily be based on outputs—discernable reactions and behaviors.

However, from a robotics perspective, the deeper question, and the more speculative one at this point, would be how far we can peer into this abyss, or even, to continue with our landscape motif, whether we could build a metaphoric bridge across the chasm of interspecies subjectivity. What would turn a performance of animal behavior into more than just going through the motions? What would make a robotic creature actually *feel* some form of reciprocal bond or kinship with us, in the manner of a dog? How could we create a model of empathy or social bonding that wasn't our own wishful projection, but was actually mutual?

The question remains speculative because there are too many unknowns—we don't know exactly what we would be trying to represent with mechanisms here. We don't know ourselves well enough, let alone what it's like to be a dog, and even the terminology struggles to define the terrain. Is "consciousness" the right term for this "hard problem" of mind-body dynamics, or is "being" more apt? Or how about "mind," as in the cybernetic intellect of Deep Mind? Could we divide the mind's functions according to some taxonomy, such as the sentience, sapience and selfhood model that has previously been put forward, or parse the brain's operations as a hierarchy of reptilian, paleomammalian and neo-mammalian layers accumulated over evolutionary time, as has also been proposed? What about "generalized human intelligence," the term Brooks uses in a series of recent blog posts to go beyond contemporary machine learning programs by including the recognition of the self in relation to the world, a self-generated sense of being. Generalized would seem a capacious enough term to include both intellect and affect, but why specify a human model? I could possess some comprehensive sense of my own existential conditions without a human, or even a mammalian, template; I could be working with the swarm intelligence of bees or a spatially distributed intelligence inspired by plants. The fact is, while we do have ample proof that this phenomenon of selfhood exists in the world, we don't have a template for it at all.

We're bushwhacking a bit here, but nevertheless there are two provisional paths marked on our biomimetic map, even if they're a bit like dotted lines sketched in pencil. The feminist theorist Donna Haraway has usefully signposted these two routes to selfhood, and we'll follow her lead here. Haraway is perhaps best known for challenging conventional boundaries between nature and technology with her "cyborg manifesto" in the late 1980s, roughly the same period as the emergence of Brooks's physically grounded creatures. More recently, however, she has tuned in to the connections between humans and other companionable "critters," a multispecies pack that often include dogs, although Haraway also mentions technological "oddkin," the kind of creatures that populate this biomimetic nature walk, even if her preferred biomimetic metaphor is the roiling fertility of a hot compost pile.

In describing the process by which the self comes into being, Haraway distinguishes between two paradigms: "autopoiesis," a process that begins with individually isolated units, and "sympoiesis," the social or interactive model she prefers. The concept of autopoiesis was articulated in the early 1970s by two Chilean neuroscientists, Humberto Maturana and Francisco Varela, who originally saw it as a distinguishing principle between the living and the nonliving: living creatures are constantly engaged in a process of regenerating and replacing cells according to internal blueprints, while simultaneously controlling the influences of the outside, and it's these mechanisms that provide the basis for selfhood. We exist as living beings because we have these self-perpetuating regulatory systems that keep our bodies chugging along in spite of everything the world throws at us.

In Haraway's critique of conventional biological research, this autopoietic paradigm reinforces a misguided emphasis on the individual as the appropriate unit of study. But given how ubiquitous this presumption is, let's consider its value. One way to understand the way a creature exists in the world is to identify the functions of all its constituent parts—to isolate the variable by focusing on the individual system within the individual animal. It's an approach that works with some biomimetic research questions: if you're trying to reproduce the gait of a dog, you focus on the mechanism that produces that motion. You could extend this approach to the nervous system: the first thing you want to know, if you want to reproduce the thinking and feeling functions of the brain, is how the brain works. How do all of its structures operate, and what roles do they play?

The engineering of an autopoietic creature would thus begin with the presumption that the mysteries of selfhood and feeling are ultimately material manifestations—they come from bodily processes, from mechanisms that have arisen in response to the evolutionary pressures of the real world. Neurobiology, the discipline that studies the physiological underpinnings of the nervous system, would be a likely place to look for such a model. We've already encountered mirror neurons, cells whose function, some theorize, is to engage with the behavior of others through imitation. Empathy is one result, but a "broken mirror," or improper function in this system, has also been suggested as a cause of autism. These cells have until recently been described in autopoietic terms. They operate, like homeostasis, at an instinctual level developed over evolutionary time via natural selection: we witness something, and this system kicks in automatically. Considered in this light, the mimetic faculty comes pre-programmed in us as individuals; it's not learned by relating to others.

The work of the neuroscientist Antonio Damasio both anticipates and broadens the implications of these neurobiological mechanisms. Damasio has written a series of books arguing that the emotions and feelings we think of as ineffable qualities of selfhood are actually grounded in the "feeling brain," in the physical structures of the brain and body and the hidden mechanisms that bind them inextricably together. Mind and body are not separate, and thus all the qualities we often consider ethereal and beyond quantitative analysis have their explanation in the structures of our physical selves.

Damasio's work offers an architecture of our emotional life, a "machinery of consciousness" based on a set of interconnected mechanisms. The subjective side of existence arises, he contends, in the constant interaction between the body and the world. Our senses are constantly gathering data about what's going on at this interface and funneling it into our nervous system. The processing of these status updates by the nervous system produces the two distinct phenomena we call emotions and feelings. Feelings, he argues, are not some transcendent out-of-body experience. They're produced by the mechanisms of the body; we might consider mirror neurons to be one of them. Indeed, Damasio's early work described a "theater of mind" in which the brain constructs a performance to generate feelings. It does so via an "as-if body loop," simulating the body's sensory experiences "*as if* they were occurring." Mirror neurons, in his view, are "the ultimate as-if body device," an extension of the mirroring we perform with our own bodies to reflect what we observe in others.

These elements of a neurobiological framework lend themselves to the design of biomimetic operating systems that go beyond cognition to include an emotional inner life. Damasio himself has recently proposed the development of robots based on another aspect of individual embodiment, homeostasis. In a medical context, homeostasis describes the way the body regulates various systems to maintain a healthy equilibrium. Our core body temperature, for example, stays within a narrow range regardless of whether we're out on a scorching afternoon or a frigid night. Our immune system identifies and battles pathogens; our pores ooze sweat to cool us when we run; our pancreas secretes insulin in response to an intake of sugar, all without our consent. To lose control of these systems, to experience a wild rise in our temperature or blood pressure or some drastic change in the chemistry of our blood, is to put our survival at risk. In "homeostatic machines" like us, these mechanisms define the boundaries of our selfhood by limiting and controlling potentially de-stabilizing interactions with the outside world.

Damasio envisions robots that are likewise attuned to their own potential vulnerability, ready to take action to regulate their exposure to the outside world and repair themselves when damaged. "Rather than up-armouring or adding raw processing power to achieve resilience, we begin the design of these robots by, paradoxically, introducing vulnerability," Damasio and his co-author, neuroscientist Kingson Man, write. In these machines, awareness of their own physical frailty would mimic the conditions that brought about the evolution of the feeling brain: self-repair would be the first step toward exhibiting emotions and feelings and ultimately the sense of selfhood that has so far eluded robotic design.

Neurobiological research is also happening on the canine side of interspecies companionship. Research published by a Harvard University team led by evolutionary neuroscientist Erin Hecht, for example, explores the neurobiology of domestication in dogs, in particular the way selective breeding has shaped the physical contours of the canine nervous system. Dogs, we know, behave differently than wolves. But how is the phenomenon of domestication reflected in the architecture of their nervous systems, and what light might those distinctive patterns shed on their evolution? In a recent article, Hecht's team describes domestication as a "massive 'natural experiment'" in neurological evolution that "has been happening right under our noses." Hecht's lab used MRI scans to compare the volume of regional brain structures in 33 different breeds of dogs, looking for the "neural underpinnings" of inherited behavioral traits between, for example, a Maltese bred for "explicit companionship" traits and a Labrador Retriever bred for "bird retrieval." They found clear distinctions in the distribution of grey matter between the breeds; the scans showed that a Yorkie's brain differs significantly from a Basset hound's in six specific regions, each containing different neurological networks and each thought to be involved in producing different behaviors. Their findings suggest that selective breeding has not just altered the shape and size of dogs, but also their brains.

This evidence of specific brain locations involved in the mechanisms of behavior also suggests the possibility of developing models that can then be translated into an artificial canine. But maybe there's something else going on here. Maybe what we're seeing on these brain scans is evidence of relationships at work, a long evolutionary history of sharing space with dogs that has shaped our own brain matter too. The way a dog comes into being via domestication is not necessarily autopoietic, in other words; maybe it's coming from the long commingling of human and canine.

It's time to backtrack a bit and start over, this time on the sympoiesis trail. Sympoiesis, or "making-with," as Haraway describes it, means creatures like Freedom and me exist in terms of each other rather than as two individuals who've already defined ourselves in isolation. This sympoietic self-making between species is in keeping with Karen Barad's reading of quantum physics we encountered in the previous chapter, in which all the living and nonliving complexity of the material world comes into existence via relationships and bonds, all the way down to the quivering bonds between atomic particles; matter doesn't pre-exist those interactions. The operative term here for being in this world is entanglement, which in quantum theory refers to seemingly indivisible states of matter, but which here also refers to creatures entangled in life webs, in ecological and social relationships that determine their existential qualities, their subjectivity, their selfhood. In sympoietic terms, domestication is a phenomenon of entanglement. "Nothing makes itself," Haraway writes. "Nothing is really autopoietic or self-organizing."

Sympoiesis "enfolds" autopoiesis in Haraway's reading; we might read the impact of selective breeding on canine neurobiology in this way, but more broadly we might think of natural selection by evolution as a sympoietic process, enfolding individual mechanisms like homeostasis and the mirror neuron system, and creating the ground conditions for selfhood.

In fact, Haraway follows the evolutionary biologist Lynn Margulis in going further, offering "symbiogenesis" as an alternative to the conventional focus on the role individual genetic mutation plays in evolution. Different symbiotic associations of microbes, like those in the human gut, interpenetrate each other on the genetic level, and in their sharing and swapping of code create the conditions for natural selection to occur. Thus, if you want to model an individual sense of selfhood, you'd have to begin by mapping the enfolding entanglement of relationships, the hot compost of various creatures all defining each other's existence. Resistance to these entanglements is futile, since the "self's thin membrane" encloses a community, not an autopoietic individual. Homeostasis, it turns out, is being performed collectively by and for all the organisms that make up the body.

The connection between mirror neurons and autism has also been questioned along these lines. The mirror we possess, these critics argue, is not biologically pre-determined in specific cells. We don't exist as beings in the world with an intact or broken mirror. Rather, research shows that even a "mature" mirror system can be altered by training. That in turn suggests our mirroring capability develops through the process of engagement with others, through social contact, through what amounts to "making with" others.

Making with, however, is not exactly the same as what happens in biomimetics, which is more of a "making as if" practice. "Making as if" does not necessarily require making with; in fact, we might discover that neither the autopoietic nor the sympoietic is definitive, that between the original and the model something resists our deterministic efforts. As with the ultimate inscrutability of Nagel's bat, the mind, swathed mysteriously in strongly emergent phenomena, resists causal analysis. From this vantage, what Dillard experiences with the mind of the weasel is not a mechanism, not something unexpected that can ultimately be explained. Her flash of insight is not a revelation in architecture or pattern;

it's not the self in isolation exactly, but it isn't making with the weasel either. She experiences herself as other. There is no hidden mechanism that a complex stochastic model can reveal. In the eyes of a dog, fleeting recognition, and the abyss.

~

When MUTANT began walking around Sony's Digital Creatures Laboratory in the mid-1990s, it was surrounded by what was, digitally speaking, an almost primordial world. No ultra-fast fiber network existed; touchscreens were unavailable, apps and the cloud were unknown. Amazon's book business was growing by leaps and bounds, but there was no Facebook yet, and making social connections on the internet had a dial-up sound, that discordant, tonal mix of twangs and beeps and onrushing static, now largely forgotten. Like Squirt the roachbot, MUTANT had to make do with a smidgeon of memory. It possessed only eight megabytes of storage, whereas today the smallest of thumb drives might offer eight *gigabytes* of memory, or a thousand times more storage capacity. It also relied on a microchip whose processing speed was at least a thousand times slower than what's inside your average smartphone. Which only makes this prototype's capabilities all the more remarkable.

In the black-and-white photo that introduced this creature to the readers of *Autonomous Robots,* the nascent scientific journal of the field, MUTANT looks like a monkey, the length and proportion of its hinged limbs making it seem as if it was built to swing from trees rather than run across open ground. With its entrails—electric motor, circuit boards, wiring and batteries—left exposed along the length of its body for ease of tinkering, it weighed in at about three pounds. Two glowing "eyes" studded the small casque of its head, anchoring facial features that resembled a black ping pong paddle, or perhaps the delicate charcoal muzzle of a squirrel monkey.

The fact that MUTANT even had a face, or a head for that matter, while its technological innards remained in full view, was an indication that this creature was being designed for an audience who assumes the face is where the action is, that the eyes are the window to the soul. We primates aggregate four of our primary senses in a surprisingly small patch of real estate, and we tend to be more kindly disposed toward creatures that do the same. In actuality, MUTANT employed a series of sensors to process what was happening in the world around it, including a microcamera that allowed it to see, a microphone so it could pick up sounds, touch sensors so it could feel the impact of someone's fingers, and a three-dimensional acceleration sensor to maintain a sense of equilibrium in space, all of which are essentially face optional. The windows to its interior workings were in actuality small and nondescript indentations, the kind of thing we humans easily overlook.

If we stopped here, with MUTANT's physical attributes, then we wouldn't be that far from HyQ and the other creatures we've seen running around various robotic labs, creatures that do what they're supposed to do and don't pretend to care how we feel about it. But MUTANT was endowed with an additional layer of componentry that set it apart, a "behavior generation system" whose goal, according to the authors of the *Autonomous Robots* article, Sony's Masahiro Fujita and Hiroaki Kitano, was to produce an "unquestionable sense of *reality*."

The goal was to make us feel that this was a real creature, by creating internal mechanisms to produce representations of emotional behavior. Just as Freedom might feel the tug of instinctual eagerness to get his nose into whatever is waiting down the street from us, and might feel anxious too about being tethered to a greenhorn like me, so MUTANT's behavior generation system was designed with an "instinct/emotion module" to influence robotic behavior. An "angry" MUTANT, for example, could act out petulance; if it was tired, it would "instinctively" seek out its charger and "eat," or curl up, chin between paws, in a classic canine pose. "Higher order" processing would kick in when the robot encountered something that required decision-making; hearing a human voice for example, would bring an additional "high-level cognition module" as well as "reactive behavior subsystems" into play, just as they might with Freedom if he had to respond to the sound of my voice, begging him to sit down.

With all of these software layers in operation, MUTANT was capable of using its camera and microphone to recognize you, and then, moving closer, karate chopping you with its robotic paw while releasing a chirpy robotic war whoop through its speakers. The goal, according to Fujita and Kitano, was to get these layers to compete with each other, so that "complex motions and actions emerge naturally," as *naturally*, I suppose, as when a kibble treat accidentally drops from my hand and Freedom has to resist the urge to snuffle it up.

We could argue that system-generated "emotions" and "instincts" aren't the same as the real thing; no matter how we label it, this is just software beneath the surface, just a complex and elaborate script, line after line of code, none of which evolved in the biotic community. But having considered the neurobiological view of behavioral mechanisms, let the significance of these terms sink in for a moment, as aspirations rather than actualities. *Behavior generation* system. *Instinct/emotion* module. *High-level cognition* module. *Reactive behavior* subsystems. From a neurobiological perspective, these designations sound a lot like mechanisms for what we call the various strands of consciousness, like what is going on in the regions of your typical mammalian brain as it navigates the challenges of daily life. Isn't this most of what an animal like Freedom, or animals like us for that matter, use to find a place to pee, or greet a friend? There are autopoietic models of mammalian brain architecture built into terminology like "higher order processing," a supposition about the way neural networks navigate between layers and across thresholds, between the primal urges of "instinct/emotion" and the intellectual pondering of "high-level cognition." We might even consider MUTANT's "mind" in terms of artificial ethology, as a model of Damasio's neurobiological mechanisms of emotion and feeling, a limited test case for the feeling brain.

But to see MUTANT in this way would mean that its "unquestionable sense of reality" wasn't just an animatronic performance meant for us as consumers—it would also raise the question of MUTANT's experience of the world.

Complicating matters for us is the fact that because the biological systems that govern emotion and instinct are not only unverified but often invisible to the naked eye, the far slope of the uncanny valley—where we can't discern the difference between the biotic and the artificial—isn't all that far off. Descartes recognized the implications of this invisibility way back in 1649: "If there were machines with the organs and shape of a monkey or of some other non-rational animal," he wrote, "we would have no way of discovering that they are not the same as these animals." Descartes might easily have been describing an encounter with the prosimian features of MUTANT, except that his purpose was not to extend consciousness to machines, but rather to deny the existence of being in animals. Since we can't tell the difference between a monkey and a machine, he argued, we might as well treat animals (excluding us humans, of course) as if they were automatons made of flesh, mindless and soulless, incapable of actually thinking, feeling, or suffering. His view that the body is a machine while the human mind is transcendent held sway for centuries; only recently have we begun to acknowledge that other species might possess all the attributes of subjectivity and deserve to be treated accordingly. The mind/body split is unfounded, recent thinking goes, as is the separation of humans from other animals. However, even as neurobiologists like Damasio assail "Descartes' error" and describe mind and body as one interconnected system, the mechanical metaphor not only persists, but ironically now extends to the human mind as well.

How does the human animal known as a nature writer engage with these questions?

Through eye contact. That is the nature writer's version of the Turing test, the famous experiment Alan Turing proposed in 1950 in which a human has to judge, based on communications from another room, whether they are in dialogue with another human or with a computer. The Turing test was blind, concealing the fact that the computerized correspondent had neither face nor eyes, but this nature writer's version is visual and utterly subjective, as idiosyncratic as Dillard's avowal that she'd been in the mind of the weasel, a biomimetic test of the entangled ontological, epistemological, and ethical dynamics of human-robot interaction.

I'm reminded of an essay by the nature writer David Quammen, exhorting the reader to look into the eyes of even the most unnerving of creatures before deciding whether to kill or let live. *Eye contact,* he urged. Make eye contact with the other. In his case, he was talking about baby black widow spiders who had colonized his writing desk, with the task of making eye contact complicated by the fact that these arachnids possessed eight eyes. Robotic creatures offer their own set of challenges. In fact, a robot-like MUTANT might seem more akin to a spider than a humanoid like SEER, and more likely to raise similar questions.

Before we get to robotic eye contact, though, let's consider a trained maker of interspecies eye contact, the service dog. When Freedom put his head in my lap and stared into my eyes for what felt like forever on our first night together, I could feel myself getting uncomfortable. Like most other species of primate, we don't necessarily enjoy unremitting eye contact; we have different durations and gradations for this gesture, and longer is not necessarily better.

What did he want? What was he waiting for? He'd been trained to keep his eye on the eyes of his human—that's how he would know if I wanted something, by looking me in the eye without pause. But all this eye contact—make it stop! Couldn't he see I was watching TV? His rubber bone was sitting on the floor, he'd just been outside…what was going on in there? There was no I've-been-inside-the-mind-of-a-service-animal epiphany. Had we spent more time together building our relationship perhaps it would've been different, but as it was I had no idea what was going on inside his head, no idea what he was trying to express or what he expected me to do. I tried to put myself in his place: still stumped. Which seemed normal; I didn't expect more. The struggle for affinity is common; the breakthrough rare. That's why we valorize it.

And yet, even if we weren't more than acquaintances really, and were reasonably far apart on the mammalian family tree at that, I could still sense that someone was in there, trying to get through to me, trying to figure me out too.

Would the same be true if I looked into the eyes of a robotic companion? Would I experience the same bafflement? The same begrudging acknowledgement of otherness? As with the eight eyes of a spider, the gesture isn't necessarily limited by the shape or location of the visual organs. When we look into a robot's visual apparatus, however, what do we see? Our own reflection, or something else?

~

There was no possibility for mutual gazing when I first met Sony's MUTANT successor, a quadruped robot named AIBO, a name that plays on the first two letters of artificial intelligence and the Japanese slang for friend. This lack of facetime came as something of a relief, actually. Looking into Freedom's eyes was exhausting enough; I wasn't ready to meet the robot's gaze right away.

"He can do tricks with his bone, or he'll chase his ball," Michael Huhns, the chair of our university's Computer Science Department, told me, gesturing toward the red plastic toolbox that serves as a robotic pet carrier. "But his behavior is not deterministic. There's no button to press if you want a specific behavior."

AIBO's box looked like the kind of thing a handyman would reach inside to grab a screwdriver or a wrench. You could fit my mother's teacup Yorkie in there, but you wouldn't do that to a dog. Even so, how easily we fell into referring to this machine as if it was a *He* rather than an *It*. Maybe Huhns was using these pronouns to make things easier for a cyber novice like me, but it felt as if our language was struggling to categorize this creature. We don't have a comfortable pronoun for a biomimetic companion robot.

Huhns popped the lid to expose the shiny black limbs of Sony's third and final generation model before a decade-long hiatus, the "sophisticated" ERS-7.

Nothing inside the box looked all that revolutionary, just a few lengths of cable, a dog-eared manual, and a plastic ball and bone. Unchewed, of course. The tangle of wires and circuit boards that made MUTANT's body look like a mad scientist experiment had been hidden away under plastic; the result looked as much like a plump edamame pod as the body of a dog.

"He'll need to be charged," Huhns said, nudging AIBO's backside to show where the batteries go, in a nondescript grey slot that nevertheless stirred a gust of nostalgia in me. I knew that battery pack slot: it was the same as the one on our old Sony video camera, which I handled all the time to take video of our kids' early antics in the days before smartphones.

What makes AIBO different from the various experimental prototypes we've seen so far is that this creature actually emerged from the lab into the marketplace. It was out there, meeting people like me, inhabiting the great indoors. Ask a robotics expert about AIBO and it won't be long before they marvel at the fact that something so sophisticated for its time was available to the general public for only 2,000 dollars. That's still a lot more than a Furby, the fuzzy little robotic owlet that came to market a year before AIBO and cost only $35. Like some giant twittering flock of starlings, millions descended on the Christmas market of 1998, inspiring a consumer frenzy that augured well for the sale of something bigger and more complex.

While Furbies made the most of a limited repertoire of creaturely behaviors, AIBO took things a step further. The operating manual that accompanied the original generation encouraged owners to take "a psychological peek" into the mind of an "independent roommate" that "was created to have emotions, instincts, learning ability and the capacity to mature." When the robot arrives it's like a newborn, the manual says; as it gains experience, it matures, first into childhood, then adolescence, and finally into an adult. This performance of robot maturation is similar to what happens in a Furby, but as Huhns told me, AIBO's maturation process isn't pre-programmed. It doesn't matter what you say to a Furby—at a certain point, it will stop speaking "Furbish" and begin sprinkling its responses with intelligible English, regardless of how much time you've invested in speaking to it. But AIBO was different: he was endowed with his own free will, and whether he turned out to be "cheerful and good" or "lazy and unruly," according to an accompanying handbook entitled "Raising AIBO," depended on how he was treated as he learned over the course of three months. AIBO learned from playing with his ball; he learned from whether we scolded him or praised him for his behavior. And according to the handbook, there was no going back—unlike a video game, AIBO could not be easily rebooted to begin the maturation process again. If he was raised to be a bad robot, he was going to stay that way.

The language here is striking, if you think about the quest for homeostatic robots and generalized artificial intelligence. A psychological peek? Emotions and desires? This is the language of selfhood and subjectivity, and indeed, the impression is furthered by the fact that the robot develops an individual, autonomous character over time, through what amounts to behavioral training. Were we to take a psychological peek at AIBO's artificial neurobiology, we'd find an algorithmic architecture—not as sophisticated and complex as those driving state-of-the art deep learning networks today, but a layered mechanism for learning nonetheless, one geared for the performance of desires and feelings. The ambition was to offer more than mere animatronics, to define parameters that would represent subjectivity and then develop mechanisms to embody them. AIBO doesn't do internet search; it won't beat you in chess. Its processes are geared to feeling rather than intellect, to learning the behaviors of the domestication process. An attempt at a feeling brain, perhaps, but without the interior stage—AIBO does not generate feelings by recapitulating and evaluating its presence in the world; it can't envision and critique its own performance. Its experiences shape its "sweet-tempered" identity, but that ability to incorporate sensory data into a lasting pattern doesn't allow it to range much further than a roachbot reacting to light or a search engine finding results in response to a query. It has no answer for the "hard problem" of the unmapped mind.

When I hear the word training, my mind immediately goes back to my training days at PAALS, which memorably included walking "loose leash" around a parking lot in a circle with eight other human trainees and canine candidates. My heart was fluttering, I have to confess, mostly out of worry that I would humiliate myself somehow, not just in front of the human trainers and trainees, but also in front of the shaggy old timers, golden retrievers and labs for the most part, who had excelled at the training and were now leading us all by their serene example. They didn't seem to acknowledge the new sounds and smells—they just plodded resolutely through the routine, furry shoulder to human hip, gaze fixed on the path ahead.

Our PAALS trainers instructed us to offer the dogs their rubber bone as a toy, but to avoid throwing them a tennis ball because the trainers used the ball as a tool to teach the dogs how to fetch items like television remotes. We should avoid doing anything that might be inconsistent and encourage unruly habits, the kind of thing that disqualifies the majority of dogs from service. We were being trained too, Just as much as the dogs, actually, if not more so.

We could say that *Raising AIBO* is a biomimetic performance of such practices, a staging of the kind of instructions you'd receive in a handbook of advice for novice pet owners. We see our own training, the practice of domestication in all its anxiety and longing, reflected there.

~

"We strongly believe that after the Gold Rush of the Internet and cyberspace, people will eagerly seek *real* objects to play with and touch," Fujita and Kitano wrote of the market for their "pet type entertainment robot," and sale of the first generation of AIBO, released in Japan in 1999 as a limited edition of 3,000 robots, seemed to validate their hopes. AIBO was greeted with must-have enthusiasm, selling out in minutes. The same thing happened within a few days in the United States. Sony produced several generations, each more interactive, autonomous and aware of its environment, until 2006, when production ceased in response to waning demand. Furbies died out around the same time, although like AIBO they've returned intermittently in more complex, and less popular, generations. It seems Fujita and Kitano's hypothesis about the value of real objects was a bit ahead of the curve; 2006 marked the explosion of interest in a phenomenon they hadn't seen coming: the iPhone, whose debut generated an entirely new virtual ecosystem. The smartphone, so tactile, so mobile, so virtual, became the real object to play with and touch. Just ask our kids.

Still, the kids were excited and nervous when I told them we were getting another dog, much as they were at the prospect of Freedom's arrival.

"A robotic dog," I cautioned them. "He's made out of plastic. We have to charge his batteries before we can play with him."

My son's first question: "Does he go to the bathroom?"

Spoken like a true ten-year-old. I was almost sorry to disappoint him—no, he doesn't. That's pretty much the whole point.

Does he have a mouth? they wanted to know. Could they pet him? They asked if he barked. If he played. If he could run around. They wanted to know how much dog there was in this thing.

"Is he big?" my son asked, with a hint of fear. I started to reassure him; this dog was about the size of his grandmother's Yorkie. But that wasn't reassuring; he was wary of Daisy.

"It doesn't bite," I told him, shifting to the objectifying pronoun. "It's just a robot."

AIBO's introduction to our household couldn't have been more different that Freedom's momentous arrival, when, as instructed, I took him on a leashed tour of the premises as soon as we arrived, and then allowed him to rest and get his bearings inside the carrier we'd set up in our dining room, his familiar home away from home. There was a lot of tail wagging, and some excited barking, and some startled and hesitant children peering around the corner at a dog—a real live dog!

The robot and I, by contrast, entered unceremoniously via the back door without a sound, no sniffing or snorting, no clack of claws on hardwood floor, just this red plastic toolbox. I thought the kids would be unimpressed, but by the time I actually got our pet entertainment robot's batteries juiced up, there was almost too much excitement in the air. The kids had that open-a-present, see-something-extraordinary, Christmas morning kind of lilt in their voices; they were dialed in on the box where this amazing addition to our lives awaited.

I was afraid I'd built this up too much; maybe AIBO wasn't all that different from the rest of the kids' toys. They already had an animatronic hamster that scurried around the floor in a clear plastic ball...if you flicked its switch. They had loons that whistled eerily if you squeezed their plush abdomens, lions that roared and shone a flashlight beam if you lifted the hinge of their mouths, birds that repeated after you in a twittery chortle.

With the power off, AIBO was just a clunky bundle of appendages; it was easiest to hold him by the back of his carapace, limbs dangling like a charbroiled lobster.

"He has no face," my daughter complained, wrinkling her nose as I held him aloft. Even her most basic teddy bears had faces. Gone were the squirrel monkey features of the prototype; this guy's mug was a disconcertingly glossy plastic lid, the kind of thing you'd pop into the microwave. His only visible feature was the pockmark of a camera, almost hidden beneath the edge of what would be a real dog's snout.

Then I pressed the power button. Twin constellations of tiny LEDS began to glow and form an eye-like pattern; they were there all along. Then the limbs began to move, stretching outward, all four paws out, in the manner of a dog awakening from a nap. The tail, which resembled two slightly curved tines of a rubbery kitchen utensil, twitched. The miniature trap door of the mouth opened with a faint whirring click, like a DVD player starting to spin. A soft puppyish yip escaped.

When AIBO stood up, the kids went crazy.

It was as if the quaint notion of spontaneous generation, of living organisms erupting from nonliving materials, frogs from swamp mud, flies from garbage, had suddenly come true. First there was a heap of plastic, and then a creature arose, right in the middle of our living room.

This thing is alive, I couldn't help thinking. *Part land crab, part puppy.*

It's this kind of experience that worries MIT psychologist Sherry Turkle, who studied the way people relate to various forms of synthetic companionship for her book *Alone Together: Why We Expect More from Technology and Less from Each Other.* "We are at a point of seeing digital objects as both creatures and machines," she writes, as "a series of fractured surfaces—pet, voice, machine, friend." Biomimetics isn't a figurative expression of entanglement; it's a fracturing, breaking what were once the distinctive categories of living creature and nonliving machine into discordant yet confusingly similar shards. Evolution hasn't equipped our brains to handle those distinctions, she argues. We may even wind up preferring the easy company of robots at the expense of our relationships with living creatures, signs of which she observed among the children in her research cohort, some of whom reported that they preferred the predictable comforts of their Furby to time with the family dog.

Even if the evolutionary stage for companionship has been set, each relationship still has to be built over time. Which is exactly what AIBO, however imperfectly, was designed to perform. What's disturbing is not the fracturing that produces more possibilities but the supplanting, the idea that the dog and the robot are rivals for our affection, and there's only room for one. They've been fused into a single category, beyond the uncanny valley, where there is no discernable difference. The robot is the dog. Is that where biomimetics takes us?

Once AIBO had wowed us by coming alive, he settled down. Way down. The manual instructed us to gently tap one of AIBO's sensors if we wanted to communicate; it said AIBO loved to communicate with us and was waiting eagerly for our attention and praise. Well, I have to say, if this AIBO loved communicating so much, he probably could've done a better job of showing it. Someone in the Computer Science department must have tinkered with his "autonomous mode," because he'd been around for years and yet still seemed to be in the latter stages of babyhood, the initial stage when he's programmed to just sit around and listen. The kids showed him the bone, they rolled the ball around in front of him, they clapped and raised their voices entreatingly, but this robotic dog wasn't running around in circles of excitement or showing off with his toys. He was timid. A bit self-absorbed, perhaps, unflappable in a way that I guess you could construe as patient. Which was actually fine with the kids—they petted the sensors in his carapace without fear. He wagged his tail and jiggled his ears, LEDs flashing signals of contentment. No scary barking, no frantic scrabbling at the end of a tight leash, but no real enthusiasm either, not the kind that had Freedom wagging from head to tail.

When the kids backed off a bit, I moved in. I got down on my hands and knees and got up close. Not too close, but close enough for it to register.

I made eye contact with the robot.

Now, to be fair, I wasn't expecting anything like the skull-splitting black holes that Annie Dillard experienced when she met the eyes of the weasel that evening by the pond. "Our eyes locked, and someone threw away the key," she wrote of that transfixing moment. What she found there was an epiphany about wildness, about freedom and instinct, and what it might mean to see the world as a small relentless predator, a transference of being. This was no Cartesian automaton.

The closer and longer I looked at AIBO's face, however, the less convincingly eye-like those configurations of lights seemed, even if "eye" was one of the words, in Japanese, that inspired the name AIBO. I recalled Freedom staring into my eyes, and this was not like that. That was Freedom making eye contact; there was volition on his part, even if he'd been trained by another human to do it. Somebody was choosing to look at me, even if I couldn't read his gaze. It was more than mutual awareness, more than sensory data being processed. It was two beings regarding each other with the long history of their species between them. Maybe we look to a dog for a sense of how a dog sees us, a canine vision of ourselves being in the world.

Although I tried to search for subjectivity in this moment, to make this face-to-face a sympoietic ritual in its own right, AIBO is not programmed to respond to someone staring into its diodes, or to seek out eye contact. There was no hint of a wild mind, no undercurrent of instinct. The biomimetic performance broke down under scrutiny. That might seem like a foregone conclusion, but then, were I to look into the eyes of a spider, would I see something different? Quammen writes of the arachnid experience:

> When I make eye contact with one, I feel a deep physical shudder of revulsion, and of fear, and of fascination. I am reminded that the human style of

face is only one accidental pattern among many, some of the others being quite drastically different. I remember that we aren't alone.

All of the above applies to me as I looked into AIBO's face, except the last. I felt alone.

"He loves Dad," my daughter said, more convinced than I was. Every time I said something, the robot tilted its head, craning its neck to catch the deeper timbre of my voice, as you might expect from a creature that had imprinted on the vocalizations of an adult back in the lab.

I kept waiting for the kids to say *this is boring,* as they eventually did with regular toys, but the fact that the robot wouldn't do everything they wanted it to, when they wanted it to, seemed to keep them enthralled. And even if they did get bored—wouldn't they have lost interest in Freedom, too?

"Bedtime," I said. "Time for AIBO to go to sleep."

I pushed the power button and he waved a paw goodbye. Down he went with a whirring of gears, just a limp bundle of plastic again. But before I could stow him away in the toolbox, my daughter dashed to her room. She came back with a small white blanket, which we folded into the box so AIBO would sleep well.

~

Meeting AIBO made me realize how much distance there is between the canine role an entertainment robot is expected to play and what a dog like Freedom is meant to do. There isn't just one model; each breed is a possible subcategory of robotic inspiration. Yet Freedom isn't defined by the neurobiology of breeding alone—that's not deterministic. Maybe he isn't inclined to conform to the rarified expectations of service; maybe sometimes he's just a dog. There are various taxonomies that seek to categorize what a dog should do or be, but the categories only partly reflect the multi-dimensional nature of two beings actually walking around together. Part of the strenuous work of training a dog like Freedom is making the dog fit the category to the exclusion of other behaviors.

The Americans with Disabilities Act, for example, codifies a functional distinction based on the work a dog performs rather than the breed. A service dog like Freedom, by definition, has been trained to perform specific tasks for someone with a disability; their role comes with legal rights to go wherever their client may go. PAALS uses a slightly more specific definition. They distinguish service dogs who assist clients with physical disabilities, autism, or PTSD from two other categories: guide dogs who work with people with vision impairment and hearing dogs trained to work with people who are deaf.

Freedom occupies a different category than a therapy dog, a dog who doesn't live with a human client but instead visits settings like hospitals or nursing homes to provide a dose of comfort and affection. And then there are the emotional support animals that a therapist may prescribe; these companions don't need to be dogs, and they don't necessarily need to be trained. They just need to be there. That's where the distinction with pets gets more than a little fuzzy.

These different lineages extend to the robotic creatures meant to perform similar tasks. Researchers describe emotional robots, for example, as a sub-set of socially assistive robots, which are in turn different from service robots. Service robots, like a robotic vacuum cleaner, are task-oriented, and they direct their autonomy and cognition toward completing specific tasks without pausing to interact with us. Socially assistive robots "mainly target the experiential aspects of belonging," which the subcategory of emotional robots provide through "interaction, communication, companionship and attachment." Other researchers distinguish between socially interactive robots and companion robots; the former refers to specific mechanisms of interactive behavior, while the latter is a function that could be fulfilled in a number of different ways, even without active engagement.

When Takanori Shibata was developing a social robot at Japan's National Institute of Advanced Industrial Science and Technology (AIST) in the early 1990s, he was looking for a model that would spark the rich and deep emotional affinity we have for our pets, without triggering the specific expectations for dog and cat behavior that would inevitably lead to disappointment. The goal would be to create a robot so convincing that it would produce the same physical and emotional feelgood factors as a visit with a therapy dog. Numerous studies have shown levels of hormones associated with wellbeing—things like serotonin and oxytocin—increase in the human bloodstream in the presence of therapy dogs, while levels of chemicals associated with stress and inflammation, like cortisol, tend to subside. Researchers have also noted a decrease in blood pressure and heart rate while participants stroke a dog and have found that our social behaviors change in a dog's company—we tend to be more friendly, more outgoing, more at ease with our fellow humans. Therapy dogs change who we are in the world, or at least those of us who are dog people.

After an initial canine prototype that test subjects found unconvincing, Shibata decided to model his robot, which he called PARO after the Japanese slang for pal, on a juvenile harp seal, a choice that makes perfect sense to me. Back in the early 1980s, when I was an avid viewer of Jacques Cousteau's undersea adventures, Greenpeace brochures began arriving in the mailbox of my childhood home. What made them so effective was the face on the cover, or more specifically the pair of eyes in that fuzzy white face: big, black, and glistening, they seemed to be brimming with vulnerability and innocence, the kind of animal magnetism that made me bug my parents for money—even if we couldn't cuddle those harp seal pups, we could at least send them all the pennies I could find at the bottom of my mom's purse.

Part of their appeal was their unattainability. These baby pinnipeds spend their early days about as far from human company as possible, out on the pack ice in the frigid waters of the far north. Most will never set eyes on a human face before they shed their white fluff and slip away into the sea. And most of us will never lay eyes on one of their kind either, unless it's on a screen of some kind, or a brochure. We have no sympoietic history with these creatures; we don't know how they're supposed to behave. As Penny Dodds, who conducted research at the University of Brighton on PARO's therapeutic benefits, told me: "Lots of people describe themselves as dog or cat people. But nobody describes themselves as a seal person."

Still, Shibata wanted to make his imitation as authentic as possible, and in 2002 he traveled to the remote Magdalen Islands, 70 miles north of Prince Edward Island in the icy Gulf of St. Lawrence, to record the sounds and behaviors of an actual seal colony. In the video of this excursion, it quickly becomes clear that baby harp seals are pretty much the antithesis of a frisky puppy; they lie around, they drowse, they wobble forward on their flippers in hapless attempts to flee. Sometimes they look up with an expression of infantile curiosity and incomprehension at whoever is holding the camera, surveying the sky and the ice, before dropping their heads once more, as if fatigued by the effort.

The sole focus of a baby seal's existence is gaining weight—they have about two weeks to triple their body weight, and they don't waste precious calories on unnecessary movement. In fact, one field study of pup behavior noted some of the plumpest specimens were so stationary that the ice beneath them melted into "cradles," like the snow beneath a mitten left out on a sunny day. About the only time they get agitated is when their mother comes back from feeding under the ice. A creature that is otherwise nearly invisible among the ridges and crusts of windblown snow can easily be overlooked, which is why baby seals attract attention by making noise, filling the air with their plaintive, haunting cries, a strategy that's probably familiar to any new parent who has fumbled through the dark to a crib in the middle of the night. Shibata incorporated his recordings of these pleading vocalizations into PARO's palette of sounds, including cries with a rising intonation which according to Dodds may imply a question mark, a sound that invites reply.

PARO looked the part of a baby seal too. Gone were AIBO's plastic exterior, the smooth futuristic contours, the sound of whirring gears. The "fractures" had been effaced, almost everything submerged beneath a coat of plush, sound dampening, antimicrobial cream fleece. Like its real counterparts, PARO has two flippers in the rear and one on either side, each driven by a motorized effector so they can wriggle independently, although they aren't strong enough for the robot to move, even in the clumsy fashion of a seal on land. Five different types of sensors, suitably disguised, pick up changes in light and recognize voice patterns as well as detecting changes in pressure, balance, and motion. Like Sony's robotic dog, PARO's software "learns" from interaction with people; it works autonomously through its decision-making protocols in response to information coming in through its sensors. It "likes" to be petted and "dislikes" the jostling of rowdy horse play. It will cry in alarm if it's held upside down, and "learns" from experience to behave in ways that elicit more stroking instead.

Most striking, however, is the face, almost an exact replica of the one on the Greenpeace brochure, so appealing, so recognizable, and yet so unfamiliar at the same time. The button nose is fringed with long black whiskers, tipped with tiny beads that improve their sensitivity to a hand brushing against them. The robot can move its head from side to side to track the movement of someone entering a room, or raise it to invite someone to stroke beneath its chin. Its mouth seems already puckered in a babyish coo. Workers at the factory in the ancient city of Toyama even trim the facial fur of each seal individually so no two robotic faces look the same, creating the appearance of individual selfhood. It's hard to keep in mind when the robot blinks as if awakening from a dream that its pair of dark and gleaming eyes, framed by lashes whose thickness and length would make a mascara model jealous, aren't actual mammalian eyeballs. The robot does possess light sensors so it has some visual perception, but that's not why it has eyes like this. They're for us to see.

I met up with Shibata, and a pair of robotic seals, at a social robotics conference at Columbia University, where he was conducting a workshop for researchers on recent developments in the field. Shibata spends a lot of time on the road, and he'd just come from a similar gathering in Nairobi, one of several stops he was making on a global tour. Having just come through airport security myself, I found myself wondering how the agents would have handled an encounter with these furry-on-the-outside, electronics-inside artifacts. I could easily imagine the jolt of consternation as the agent tasked with watching the x-ray screen saw these guys come through the machine, the frantic wave for another set of eyes. Would PARO require a separate grey bin? Would a brief petting session next to the X-ray machines, the equivalent of turning your laptop on, suffice?

Shibata seemed surprisingly unruffled, given the time changes and the rigors of air travel. Speaking English, his hands often massaged the air as he sought to conjure a word, then swept down to smooth the front of his black polo shirt. Unless he was holding PARO that is, in which case he cradled the seal in the proper manner, with one arm tucked beneath its body and the robot's face gazing out calmly from just beneath his chin.

We spent our time between sessions at the PARO demonstration table, where Shibata stroked the seals with long, practiced strokes while the other participants milled around us with cups of coffee, glancing surreptitiously at name badges.

"PARO is meant to be picked up and held close," he said, turning to show me the robot's batting eyes and bobbing chin.

"This is how PARO differs from AIBO."

It seemed to me that the baby seal would be a great panacea for the awkwardness of the occasion—how much better if all these engineers and social scientists could just pass PARO around to break the ice. Shibata himself was certainly approachable, and yet nobody seemed to want to approach the robots; the pair idled on their card table like it was a sheet of remote pack ice. It was as if the knowledge that these were therapeutic devices meant that touching them, or appearing to use them for our own benefit, was somehow inappropriate. It felt like the antithesis of Haraway's "making with" was going on here, but in actuality it was an indication that sympoietic engagement is not a simple matter. People engage with these creatures in a variety of contexts, and they bring different expectations to the encounter, just as they would with a seal in the wild.

In designing the robot, Shibata had anticipated these different reactions by incorporating what has been described as "interpretive flexibility;" not only are the seal's behaviors not deterministic, but they're subtle enough to be open to conjecture. Each culture views PARO differently, he told me, as if aware of the subcultural dynamics going on around us. In Japan, he said, people don't have much experience with therapy animals, but they're comfortable around robots. As a result, most PARO in Japan have been purchased by individuals to provide companionship rather than therapy. In Europe, the opposite is true; people there tend to believe in animal therapy, but they're less comfortable with robots entering their personal lives. PARO is widely used in therapeutic settings in Denmark, for example, but almost unheard of as a household device there.

Attitudes in the United States fall somewhere in the middle. The robot achieved something of a milestone in 2009, when it was designated as a Class II medical device by the Food and Drug Administration, a status it shares with the majority of medical equipment for sale in this country, including powered wheelchairs, pregnancy kits and nebulizers. The approval was based on a "reasonable assurance" of the safety and effectiveness of the device; enough evidence had accumulated on both counts that PARO was exempted from the pre-market testing that might otherwise be required.

Perhaps the most frequently cited study of PARO's therapeutic impact was carried out by researchers at the University of Auckland. Forty residents of an elder care home were randomly assigned to two groups, one of which had a weekly session with a therapy dog as well as with PARO, while the control group had sessions with and without the dog, but never with PARO. Researchers conducted baseline interviews to get a sense of the emotional wellbeing of the subjects before the sessions began, and then observed how many times people stroked or interacted with PARO or the dog in each group. After several weeks of sessions, they interviewed the participants again about their emotional state.

Both the dog and the robot had a significant positive impact on the emotional state of the participants, the researchers found. But the residents stroked or talked to the robot more than the resident dog. They talked to each other more about the robot too—the robot stimulated social interactions between people to a degree that the familiar pooch did not, leading the authors to conclude that PARO "may be able to address some of the unmet needs of older people that a resident animal may not, particularly relating to loneliness."

Was it just an emotionally unavailable dog? Maybe not. Some of these benefits may actually come from the way the staff facilitates the sessions with this "interactive third," as an emotional robot has been described, a point made in a talk at the conference by Indiana University's Selma Sabanovic, whose research in similar settings concluded that PARO worked best when a caregiver encouraged the interaction between the robot and the elderly patient. Which is a formal way of saying that PARO on its own is not the point; the robot comes into being by engaging with clients and caregivers as they interact with each other. It's a sympoietic device.

I'd witnessed a similar dynamic a few months earlier when I visited The Burrowes, a residential facility for clients with dementia that had participated in Penny Dodd's research, set among the sheep-dotted downs above the city of Brighton, on England's south coast. Dodds arrived with PARO in a sturdy black carrying case that looked like it might hold a musical instrument. Dodds has broad shoulders, deep set eyes and a rock star's short, tousled blond hair; in the car she was funny and talkative, but after we sat down in the visiting area outside the doors that led to the actual facility, she seemed to melt silently into the background as a trained observer is meant to do. Quite a large and expectant audience formed for the two women who came out to demonstrate PARO's therapeutic potential: in addition to Dodds and me, several of the staff nurses, therapists and interns joined us in a circle of low plastic chairs, with two robots on the table between us and our cups of afternoon tea.

Although this secure facility cares for clients with the potential for challenging or violent outbursts, everyone was on their best behavior, the women dressed as if they were going out, in dresses and jackets that seemed a couple sizes too large for them now, bulky handbags in their laps, their silver hair smoothly brushed. They were clearly delighted to be the center of attention, enjoying what Dodds described as the "halo effect of celebrity and novelty," all our eyes fixed on what they'd say or do next. But it was only with the subtle insistence from one of the nursing staff named Philippa who clearly knew what she was doing, that each of the women took turns interacting with the robot in the spotlight of our attention.

"Isn't he a good boy," a woman named "Dawn" repeated as she tickled PARO's chin, her large blue eyes twinkling with mischievous amusement. We all nodded in agreement.

"My friends all think I'm crackers," she added to no one in particular, a refrain she returned to frequently. It made me think about how neurobiological research is often focused on identifying cognitive or emotional deficits, on what's missing at the individual level, while the action here was focused on what we could create together, between us, through conversation. But if the robot is a tool for learning about human nature, part of what it reveals is mystery, the enduring mystery of what it means to be human.

Philippa asked if Dawn thought PARO was a girl or a boy.

Dawn considered briefly.

"A boy."

"Why?"

"Because he's just…"

Before the pause could become perplexing, Philippa jumped back in.

"What would you name him?"

"I'd have to get to know him," Dawn replied, looking slightly exasperated.

"What would you say if I told you this is a robot?"

Dawn wrinkled her nose, as if she'd suddenly been plunged into Mori's uncanny valley.

"I'd be shaken a bit," she replied. "I'd have never thought you'd say that."

Philippa pushed further, turning PARO over to expose a seam running along his abdomen.

"Would you like to see what's inside his zipper?"

Dawn nodded, eyes widening.

"Oh yes!"

Philippa unzipped the fur to expose the rectangular plastic plate within.

"There's the battery, see?"

"Oh, yes," Dawn nodded, then shrugged.

Philippa sealed up the belly and rolled the seal over again.

None of this could happen with a therapy dog, I thought.

Shifting gears, Philippa asked whether Dawn's grandchildren would like PARO. Dawn responded with a story from her childhood. They had cats when she was young, she said, and her mother was always wondering where the cat was.

"'Where's that blasted cat then? Better not be in the bed!'" her mother would say.

"And that's right where he'd be, with us."

We all shared a laugh. And then we put PARO away together in the black plastic carrying case. Before we closed the lid, Dodds showed Dawn the bright pink pacifier that fits in PARO's mouth and charges his batteries, to reassure her that he would be comfortable.

This session with Dawn felt a bit staged, but that staginess is the point of an interactive third; it's what draws the participants into engaging with each other. Every time you bring PARO out, you're staging an encounter with an animal. It's a performance, and thinking of it that way, as robotic and human actors performing their roles in a semi-improvisational play, actually helps it succeed.

~

Another way to describe PARO's role is by describing its habitat, the ecology of the space in which it's meant to function. That's clearly not a professional conference; instead, PARO is often part of what has been called the "ecology of aging," an "ecosystem" defined by relationships between people, artifacts, and space. It's a sympoietic landscape that PARO is designed to occupy.

I wanted to venture further into this terrain, to literally walk around with PARO rather than just observe. I still found myself wondering how this emotional robot fit into what was happening in the midst of an actual day. I know enough about long-term care to know that the visiting area and the actual floor of a secure facility are two very different environments, mostly because my mother worked in nursing homes for her entire career, starting as a staff nurse and working her way up to director of nursing, overseeing care for the entire facility. She came home with plenty of stories—something was always happening it seemed, and I had a vicarious impression of these places as hotbeds of activity. The nursing home, if things are working to plan, is a highly orchestrated series of daily rituals and routines, and it was my mother's job to be the conductor. At the end of the day, she would collapse on the couch, prop her feet up on the coffee table, and announce with a marathoner's sigh that this was the first time she'd sat down since she'd left this morning. She'd had lunch on her feet, rushing from floor to floor, meeting with housekeeping, supervising the stocking of the med carts and the bathing routine, making sure everybody was where they were supposed to be and doing everything according to individual care plans.

Thom Strickland, the Activities Director at the Bentley Adult Day Health Center in a leafy neighborhood of Athens, Georgia, home to the state's flagship university, was gracious enough to invite me to join PARO's rounds for the day. The center serves a diverse population: among the 45 adults who gather here, many are elderly, some with Alzheimer's or other forms of dementia, but clients with various other medical conditions take part in activities here too.

After buzzing through the locked front doors, I was met by Jennifer, an effervescent young woman who would serve as my unofficial guide.

"They be lovin' on him all morning," she told me, making a cuddling motion, her fingers hovering over the joystick on the arm of her power wheelchair.

"*My baby,* that's what I call him."

While Bentley, as the center's robotic seal was known, was charging up via the hot pink pacifier, I followed Jennifer in a circuit of the corridors around the main dining and social area, which she navigated deftly, turning, pausing, introducing me to people.

Some elderly clients sat nearly motionless; they had to be helped up from their chair, propped step by slow step to the bathroom. For two women the opposite was true; dementia had made them hypermobile, so restless they couldn't sit down. They paced the corridors, location sensors dangling from a safety pin between their shoulders where they couldn't reach them. When they approached the exits an alarm sounded, a wailing that cut through our conversations again and again.

One of these women was deceptively frail; she seemed to float on her toes, almost unnoticed, smiling pleasantly as she tried to push open the doors and keep going. Jennifer introduced us and we stood briefly, our eyes meeting for a fleeting moment. I was reminded of the whole notion of external cues offering only an inkling of the subjective interior world of the self. The abyss between species, yes, but also between us. I couldn't quite accept there was anything different. I expected certain rituals, and we performed them impeccably, without revealing anything.

Don't be fooled, two sturdy women in cobalt scrubs cautioned me half-jokingly, as if reading my thoughts. They must've known how easy it would be for me to obligingly hold open the door for this woman, since to me she might just as easily have been one of the staff, bustling toward another task. I watched her disappear around a corner, and before I knew it she reappeared from the other side, making a circuit, then stopping abruptly to rattle the door handles. The alarm sounded. She smiled warmly as she passed by again.

Once Bentley was sufficiently charged up, Thom took us out on the floor, a bright cafeteria-like space with tables looking out through a glass curtain wall to a lovely courtyard with climbing roses in bloom. He pressed the button, hidden under Bentley's fur, to get him up and running. Or should I say, up and blinking, and maybe quivering a bit as his chin rose to survey the room. We proffered him to a woman whose avian features were pinched in a permanent scowl; she gave him a pat and pointed an accusatory finger at us, as if she'd caught us red-handed in some kind of hooliganism. Meanwhile, Bentley offered a kind of plaintive and inoffensive bleat, the kind of sound an inquisitive cat might make when saying hello to a stranger. A smile broke through; the woman's hand dropped to her side. But that was enough robot for her; she didn't want him on her lap.

After a few of these visits, Thom cradled PARO toward me.

"Would you like to hold him?"

We approached a round table with three elderly men sitting quietly. Thom made the introductions, and I took my cue from him.

"Would you like to hold him?"

I could feel Bentley's head rocking slightly beneath my chin, his fur brushing my neck.

The man nearest me had rheumy eyes beneath a flat cap.

"No, no," he said in a faint, raspy voice, lifting a hand.

"He's afraid of dogs," said his neighbor. "He had a bad experience with dogs."

"Oh, well you know, it's actually a seal," I said, trying a different tack. "Bentley. He's a robot."

"No, no," said the man, still smiling.

At most of the tables, the people were glad enough to see us; they were willing to offer a polite pat or two, but they were not exactly eager to take Bentley off my hands. And at six pounds, Bentley, who is supposed to feel comfortingly substantial in the lap of someone who is sitting down, started to feel authentically warm and hefty. I found myself fantasizing about tucking him under my armpit, head poking out like some Beverley Hills poodle. Or what about putting him in one of those front-facing baby carriers, so I could walk around with him strapped to my chest, waving his flippers at passersby?

We had plenty of time to make eye contact, Bentley and I. He looked up at me as we made the rounds. I looked down at him. He blinked, slowly, like a drowsy infant. His eyes were uniformly dark, like Dillard's weasel's eyes, perhaps. I thought back to Freedom holding my gaze—and my discomfort. The fact that he didn't seem to blink—that made me nervous. And thinking back, maybe his eyes moved in tandem with mine, searching for connection, a minute quivering and contracting, tiny but perceptible gradations that this generation of robotic seal doesn't mimic.

But if it did, what then? What would that gaze be like? I imagined it in comparison to my own defining moment of eye contact. As a nature writer, of course I have one. I was working for the Forest Service near Augusta, Montana, conducting surveys of neotropical migratory songbirds. We would gear up before the crack of dawn, jump out of our tents with binoculars and clipboard and hasten to the start of day's transect—we had to be there within half an hour of the dawn chorus, and there was no time for coffee or any lazy morning rituals. Half asleep in the chilly dark at the start of the transect, we would shiver as we recorded what we saw and heard for ten minutes, then move down the trail to the next point. The sun would gradually rise over the ridges, gilding the tops of the conifers. Warmth would spread down to our faces, then our hands.

On this particular transect nothing appeared remarkable. I walked through a shaded patch of willow on my way to the next stop, and the trail ahead ascended over a scree slope where the sunlight beckoned. As I trudged along I heard a commotion in the willows in front of me, which I barely acknowledged.

I'd been doing these surveys long enough that having a mule deer, or even an elk or a moose bound across my path in the morning was nothing noteworthy. I usually just went about my work; the other creatures went on their way.

I assumed it was a deer. My head was empty of thought. My job was to identify birds; sometimes I raised the binoculars and looked for them, but they were busy looking for each other high in the canopy. Mostly, I just listened. I noted the trill of a chipping sparrow here, a junco there. Mountain chickadee calls. Nothing unusual.

Then a figure broke from the willows and loped up the trail in front of me. It paused at a rock outcropping, staring down at me.

Our eyes met. I felt as if paralyzed mid-step, broken out of the usual trance but unable to move.

"That's a wolf," I heard myself exclaim, although there were no other humans for miles. It was so unlikely an encounter, given that these wolves were the only pack in the state outside of Yellowstone at the time. I wasn't looking for them. I had no preconceptions.

We saw each other. That's the simplest and most accurate way to put it. Across the abyss. I had no sense of my own category in that moment; I was a creature out in the forest, not on my own turf, not setting the terms of my own definition. At the same time, I never felt I had entered the mind of the wolf; it was a mutual acknowledgement at a distance. We existed in the moment, but apart, observing a kind of etiquette I suppose.

Creatures in the wild have their own business. The wolf had her things to do. I had mine. I could see movement and shadow on the other side of the rocks—the rest of the Augusta pack, hidden from view. The wolf broke eye contact and disappeared behind the rocks. By the time I got up there, slowly, not wanting to intrude—they were gone.

~

Thom and I eventually left PARO on a table, just to the side of a woman in a denim baseball cap whose long silver-tinged hair had been carefully combed over her shoulders. She'd been sitting alone all morning, seemingly indifferent to other people, cloaked in silence. She refused to respond to Thom's cajoling, gazing stonily off into the decorative white flags hanging from the rafters.

It's her first day, Thom whispered to me as we walked away. *She doesn't want to be here.*

For quite a while, nothing happened with Bentley as I sat in the corner and observed. It was incredibly loud; a musician had arrived and was pounding out show tunes on the piano in the corner, and the door alarm kept going off. The staff amplified their voices so they could get through to people with hearing aids; chairs scraped against the floor as aides helped people up for the trip to the bathroom. I could see what my mother was talking about; I could imagine her here. And yet there was a stillness in the air at the same time, so many motionless people in the room that it felt as if I could be looking at a photograph.

As I watched, my phone began to vibrate. It had been in my pocket the whole time, on silent mode yet seeking attention, nevertheless. Eventually, I couldn't resist. I stood up and walked into the hall to chat briefly with Nicola about the kids.

When I returned, I found Bentley still sitting on the table next to a furled newspaper and a puzzle magazine. Motionless. It was hard for me to pay attention, to enter into this way of being, this experience of time passing. I spaced out, thinking about something, my mind moving around even if my body sat still. I was outside, wondering what kind of roses those were, inhaling to see if they were scented, thinking about the drive home. I wondered if we really want the robot to be devoid of any biotic references, to evolve into a distinctly technological, unnatural creature, eyeless and faceless, like the phone. My mind wandered to Freedom, how he lay still during the training lectures, parked at my feet, and then grew restless at the sound of a jingled collar, the thump of a tail. How he leapt from the travel kennel when we returned to PAALS from his visit to our house, domesticated but still unfathomable to me. How would he have coped with this unnerving stillness?

When I looked again, I saw a hand resting behind Bentley's ears. The woman who didn't want to be here had reached out, her parchment fingers stroking the robot's fur. And Bentley appeared to have arched his neck, as if he was enjoying it.

5

TRAIL'S END

Burgers and biomimetic bodies

Our big family adventure in carnivore studies began on a plush new jet heading north from London, glaciated peaks and glimmering seas beneath us, lit by the bright twilight of the Arctic sun. The kids were deeply engaged—in their gadgets, my son playing a World Cup edition of *FIFA* on the iPad, my daughter battling zombies with plants on the iPhone. *Bok Choy!* she kept muttering, stabbing the screen with her little index fingers, while Brook swiped his way down field. We tried to get them to look out the window by prodding them and exclaiming gaily, "Look out the window!" But it's a four-hour flight, after all, to Tromsø, an island city north of the Arctic Circle, and they were still getting over the ten-hour, up-all-night, way-too-much stimuli jaunt across the Atlantic. Push too hard, and they'd start to cry. Or we would. Jet lag seems to hit parents harder, like a ton of bricks, basically.

From above, northern Norway looked like a primeval world. No rectangular geometry of crop fields and roads down there, none of the circuitry of residential development. Somewhere in those steep green folds were herds of domesticated reindeer, real live versions of the creatures most of us know only through Rudolph, a Montgomery Ward catalog creation from the 1930s, or Disney's more recent version, the companionable Sven of *Frozen* fame. Down there too were the indigenous Sámi reindeer herders whose lives are intertwined with the seasonal rhythms of the animals they care for … and consume. I felt as if we were not only above the ground but traveling through turbulent currents of history; looking down, it was easy to imagine the moment when wild animals first began to associate with humans, but look up and there was the illuminated sign for free onboard Wi-Fi, a reminder that the Sámi live in the 21st century too. The relationship with these animals isn't frozen in time; it's still evolving.

We were headed to a place where a carnivorous diet is close to compulsory, if you're going to eat what tundra and taiga have to offer. Before we left, I read something in the journal *EcoHealth* that I didn't share with the kids: "Some estimates suggest that reindeer can be exposed to attacks of approximately 8,000 mosquitoes per hour during the mass appearance of blood sucking insects, which locally is known as a 'rakka.'" Eight *thousand* attacks per hour? More than two bites per second? I'd been bitten by plenty of mosquitoes, but I'd never been to a place where their abundance deserved a special name. The kids were going to have a blast with that.

Why expose them to this bite fest? Why donate your children's hemoglobin-rich corpuscles to the arctic ecological economy? In my defense, let me say that our kids had never met a real reindeer. That's important, because it epitomizes the experience most of us have with our meat—we meet it in the package or on the plate, but almost never on the hoof. Industrial farming is the status quo, and in our household we live that reality every time we go to the supermarket, or drive past the urgent flotsam of fast food outlets that runs the length of the road out of town, or even when we pass the chicken plant and see, out of the corner of an eye, a scrim of white feathers floating on the tarmac. The conventional production of animal protein is a vast, dirty, open secret; how easy it is to think of meat as simply a product, while forgetting the process that made it available.

This chapter is about that process and the practices of meat production as much as it is about the product, and it takes us down two seemingly divergent paths. One, venerable and fetishized but rarely trodden these days, ventures through artisanal and locavorish variations on Wendell Berry's Kentucky farm, then veers off toward the wild and remote corners of the planet, the kinds of places where a dietary inkling remains of what it might mean, in the words of the deep ecologist Paul Shephard, "to go back to nature." I tend to think of this route as a renunciation of technological progress, although the reindeer herders zooming around on snowmobiles might have something to say about that.

The other path heads for the laboratory, where "techno-optimists" hope to turn the factory farm into an agricultural version of the eight-track tape. So far on this journey, we've been witnessing the transformation of biomimetic tools, or artifacts, into agents with various degrees of autonomy, intellect, sensory awareness, and even a foretaste of selfhood. But we've reached the signpost for an alternate route here, one that doglegs sharply into the ethically rocky terrain of subtraction, the elimination of all the features that might enable autonomy or self-awareness. The path, in other words, of transforming agents back into artifacts.

Why would we want to do that? We can't eat hardware and software, but what if we could use biomimetic technology to extricate the useful body from the sensory and cognitive capabilities of the animal mind? What about reverse engi neering the flesh of a cow, mimicking beef with genetic modification techniques, so that the flavorful proteins make it feel *as if* you're eating beef but the source is microbial rather than bovine? The whole point of these techniques is that they take biomimetics a step further than you might find in your typical veggie burger. These aren't just rough facsimiles of the real thing; they're reinterpretations of the practices of meat production, performed at the cellular and molecular level.

To a degree, it all sounds like the modern meat eater's fantasy come true: dusty feedlots and battery cages disappear, with profound implications for the planet's ecology and climate. In their place, "meat" appears as if through some form of clinical alchemy, rather than through the messy and disturbing process of raising and slaughtering vast herds of livestock. No risk of fecal contamination and superbugs with this "clean" meat. It's as if we're consuming the flesh of an animal, but it isn't literally the same as the original model. It's a performance of animal flesh, one with benefits and consequences.

This journey into robot country has asked a lot of us—to observe and touch these creatures, to walk with them and imagine how they might behave in the wild. We began with a new paradigm of plant intelligence, a sign of a broader movement to extend subjectivity to those considered inert or objectified in the past, including animals that might once have been compared to machines. The general idea was to make machines more closely resemble these complex self-aware beings as models. Now, however, we're heading in the opposite direction, doubling back into Cartesian objectification while at the same time retaining the sense of a biotic metaphor. Not of a cow grazing placidly in a field, however, but of a cow's disarticulated flesh. We're perfecting a biomimetic reproduction of an already objectified and abstracted model, in other words. Biomimetics in this context actually alludes to what we might want to forget—that beef comes from cows. In conventional agriculture, there's a whole process of transforming the animal subject into an object that lies hidden by design. Nevertheless, that's the model here, for better or worse.

There's something about tasting the flesh of our technology that feels different from an encounter with even the most unsettling of robots. The apparatus is no longer out there, doing its thing with varying degrees of intelligence and autonomy. It's not something we ultimately experience as other. It's more intimate than that, deeper even than companionship. It's intended to become part of our bodies—to be incorporated into us. It's the recognition of our own embodiment that makes the metaphor hard to swallow.

Where should we go in search of a meal, toward the hunters and herders, or toward the bright clean production plants of cellular agriculture? That is the 21st-century carnivore's conundrum.

~

I don't know why I thought Tromsø would be teeming with *Rangifer tarandus*. I half expected droves grazing the lush green margins of the airport; I envisioned a domesticated version of the epic migrations of Alaskan caribou you see on all the nature channels, the vast herds flowing like a river across the tundra, except here you could sit and watch them go past while sipping an eight-dollar cup of exquisitely prepared coffee in the "Paris of the North." Must be the magnification effect of the internet; after you spend hours online scrolling through video clips and blogs and websites, you begin to think this virtual smorgasbord of promotional material is what's actually out there.

In reality, about 87,000 square miles of northern territory, or about 40% of the country, is designated for reindeer to forage, calve and most of all move. Like their wild Alaskan counterparts, domesticated reindeer are migratory animals; their seasonal journey threads a line of survival between extremes, skirting the coast's wet and unpredictable winters by heading inland, then fleeing the interior's hordes of biting insects in the summer by returning to the coast. For some, the ancestral route is both terrestrial and aquatic: their summer range lies offshore, in the constellation of islands close to the coast, so near and yet so far. Reindeer can swim, craning their necks, following a bold and restless leader who first takes the plunge. But they can also drown, victims of a sudden storm, stiff headwinds, churning waves. Most of the Tromsø region's population now migrates through the city by boat; a Coast Guard ferry picks them up at designated points and carts them across the fjord.

Back in April, if we'd pulled the kids from school, we might have watched the reindeer ferry pass right under the bridge that connects Tromsø to the mainland. We could have stood on the shore and waved, and it would have been cold enough for the fleecy hats and coats and mittens we've stuffed into our biggest duffel bag. But summer is a different story. It was pushing 90 degrees on the city's treeless streets now, more tropic than arctic, and the reindeer had long since melted away into the mountains.

It felt bad enough being a human with no refuge from this record-breaking heat, and then I thought about what the reindeer must be experiencing. Swaddled in a coat that features up to 2,000 hollow, warmth-trapping hairs per square centimeter, the true marvels of these animals' physiognomy have been sculpted by biting winds and subzero temperatures. They're *chinophiles*, to get technical about it, creatures adapted to the snow. Whether it's what the Sámi call the *guoldu*, "the cloud of snow which blows up from the ground when there is a hard frost without very much wind," or the *goahpálat*, "the kind of snowstorm in which the snow falls thickly and sticks to things," they're ready. Like some waterfowl, they can constrict the blood flow down to their legs, which are mostly tendon and bone, allowing their limbs to drop to just above freezing while preserving warmth in their core. Just as bats have evolved noseleaves to suit their echolocation needs, reindeer nostrils and lungs have evolved baffles and chambers to draw warm moist air into the body and keep freezing dry air out. Even their eye structure is uniquely adapted to deal with winter, changing from summer gold to icy blue in order to absorb more light during the long months of perpetual darkness. None of these specialized adaptations help much, however, when it's sizzling outside.

As "evening" arrived, we settled into the family room on the third floor of a small hotel with sweeping views of a park and the downtown port beyond, but no blackout curtains or air conditioning. The sun never quite sets at this time of year, so it was bright as a South Carolina afternoon all night long. Midnight came and went, and we couldn't keep our eyes open, couldn't keep them closed. The kids couldn't wind down. An oystercatcher chattered from the eaves above our open window; I looked out to see it waddling along the edge of the road, pecking at the gravel. Down the hill, Himalayan blue poppies nodded in the twilight breeze; I'd never seen them outside of a gardening magazine before. They were turquoise, the color of icebergs. It felt like I was hallucinating, like we'd signed up for some kind of sadistic experiment in sleep deprivation.

By the time we headed south from the city, curving along the bays and inlets of the Balsfjord, a narrow channel of saltwater hooked like a thumb around the violet-tinged peaks of the Lyngen Alps, I was beginning to wonder if we'd ever see a living specimen of these mythic livestock. Our route was lined with small farms, red barns and hay fields hugging the banks of the fjord, evidence of a climatic anomaly that begins not far from our own coastline back home. The Gulf Stream, invisible to the naked eye but red as a serpent's tongue on an infrared meteorological map, unfurls north from Florida, surging past the barrier islands and the rich, manatee-haunted marshes of our own South Carolina coast. After it crosses the Atlantic from Newfoundland it forks, one tine curling toward Africa, the other arriving here, still bearing the heat of the tropics. Norwegian farmers once split their time between land and sea in the resulting microclimate, grazing hardy breeds of dairy cattle and fishing for cod, straddling the boundary between the wild and the tame. Their barns, often situated right next to a small dock, are the last signposts of the agricultural juggernaut that began with the domestication of cattle.

Signs of a shift from hunting reindeer to herding them begin to appear in the archaeological record about 3,000 years ago, although there are less definitive indications that herding may have begun as early as 7,000 years ago. Some of the earliest evidence can be found just to the north of Balsfjord, near the traditional Sámi trading center of Alta, where drawings carved into rock depict a herd of reindeer surrounded by what appears to be a circular corral. Inside, a human figure raises a stick overhead with both arms, as if guiding the animals. A century before the first Norse settlers embarked for Greenland, the Tromsø-based chieftain named Ottar boasted to the Anglo-Saxon King Alfred that he owned 600 reindeer, including six highly prized decoys used to lure wild deer into the fold. The anonymous courtier who recorded Ottar's accomplishments for posterity, however, wasn't terribly impressed. "He was among the chief men in that country," this scribe reports in one of two preserved copies of the Old English tome, "but he had not more than twenty cattle, twenty sheep and twenty pigs, and the little that he ploughed he ploughed with horses."

On our way through the bucolic landscape that was once Ottar's home turf, we failed to encounter a single grazing cow. Industrial scale farming has led to consolidation in Norway's dairy industry, and like the Norse settlements of yore, many of these small and remote operations have been abandoned, leaving empty barns and fallow fields. It says something about the ebb and flow of animal husbandry in our culture, the way an animal that seems to lie at the heart of who we are can still become marginal over time, even as the fantasy lives on.

We didn't see any reindeer among the farmsteads either, but we did pass a cluster of *laavos*, traditional round dwellings of stripped poles wrapped in fabric, repurposed here for the tourist trade. Inside the first, a birch log fire smoldered beneath a scorched aluminum pot, the contents of which were dark brown and porridge-thick, the skin on the surface pocked by the occasional bubble. A small hand painted sign hung nearby.

Reindeer stew, it said, in English.

Next door, surrounded by plastic woodland gnomes and other tchotchkes, was a man with puckish features framed by a sweep of brown hair. *No English,* he said shyly, rubbing his brow with the back of his hand as I tried to suggest that we were looking to try the local specialties—but not that particular batch of stew.

He understood; yes, his sister knew someone who ran a traditional restaurant near Alta; he'd call her and…

Suddenly, I sensed a small presence at my side.

"No!"

The last time I saw her, Beatrice had been fingering the flags and mittens and reindeer antler keepsakes, looking for something *Sven*-like. I assumed she was out of earshot, but suddenly she was right at my elbow, scowling as the man took out his phone.

"Sorry!" I said, smiling inanely and trying to pretend I hadn't heard my daughter.

"I don't want to eat reindeer," she declared firmly.

I knew where this was going, but I tried to feign amusement, nevertheless.

"It's not just a stew. It's a traditional meal."

"No!"

She was getting louder. The man's finger hesitated over the numbers.

"There's singing. Everybody sings fun songs."

"No!"

Louder still. Should the man dial, or…?

"All right," I said, "We'll talk about it in the car."

"No!"

Even louder.

"Why don't you go see what Brook is doing?"

"I DON'T WANT TO EAT REINDEER!"

Cue the soaring vocals of *Let It Go*.

~

Eventually, we reached Nordre Hestnes Gård, a formerly derelict small farm trying to make a go of it on the steep edge of the fjord. With the exception of the thoroughly modernized guest apartment, the operation hadn't been gussied up for visitors like us. The farmhouse was an unprepossessing place with faded yellow siding and a collection of tractor parts and various odds and ends in the open bay under the barn. The farmers were hard at work when we pulled in the drive; Bård was on a tractor, layering hay to dry in the narrow pasture across theroad, and Gry was on the steep hill above the barn, spreading some white shade cloth over the vegetable patch while a sprinkler jetted an arc of water across the rows.

They'd been working to restore this place for years now, renewing the fields and reintroducing livestock. No cows, but they had Icelandic sheep grazing on the hill and Icelandic horses in the barn, some pigs foraging in a patch of woods, lots of chickens. We had to shoo away a flock of hens and Muscovy ducks from the shady doorstep of our apartment, which the farmers had recently converted from the front third of the barn. We couldn't get much closer to living with domesticated animals; the henhouse was literally right on the other side of our bedroom wall, everything on our side immaculate and modern, everything on the poultry side chalky and pungent. Beyond the chickens' abode were the horse stalls, shadowy floors clotted with hay and manure. There was fresh white poop all over our front step.

Just what I was hoping for.

Our kids looked stunned, cowed into wide-eyed silence by everything going on around us. Animals seemed to be everywhere: lambs bleating, dogs barking, chickens clucking. A trio of stocky, chestnut-hued horses stood at the corner of the barn, twitching away flies. We heard the occasional thud of their hooves, heard a cat, shut inside to recover from a recent spaying, yowling plaintively at a window. Interlaced with the animal sounds was the familiar back-and-forth of human voices, but in a language we couldn't understand.

Once we'd unpacked a bit, we watched as one of the couple's teenage daughters and her friends rode past on the horses, trailed by the tousle-haired youngest daughter on her Shetland pony Dauphin. One by one, the bareback horses ambled across the stony beach and waded out into the clear calm water of the fjord. The girls splashed and whooped while we looked on. It was like something out of an old film, archival footage of a long-lost childhood. The iPad was tucked away somewhere; nobody had asked for it in quite some time.

There were reindeer on this farm too, but they didn't belong to the farmers. According to Norwegian law, only those Sámi whose parents or grandparents were reindeer herders have the right to herd reindeer today. This exclusive recognition reverses a long stretch of discrimination and forced assimilation, including a span of decades before World War II when the Norwegian government essentially tried to stamp out Sámi culture by forbidding their language, denying their land claims, and even sterilizing some Sámi women. However, recent policy changes, partly in response to pressure from Sámi activists, have granted the Sámi greater autonomy and control over local natural resources. A separate Sámi Parliament was established in 1987, in Karasjok, a small city in the middle of the government established Sámi Reindeer Herding Area. A year later, an amendment to the constitution acknowledged "the responsibility of the authorities of the State to create the conditions enabling the Sámi people to preserve and develop their language, culture and way of life." Updates to these policies have shifted reindeer herding toward joint management arrangements with Sámi herders, right down to the local level, where the traditional *siida*, or communal herding group, is now the recognized authority.

Bård said one *siida* kept a token herd of about 20 head in the mountains above the farm. In the spring, the herd wanders down through the farm's pastures to eat seaweed and lick salt off the rocks by the fjord. Back in April, he said, you could see them practically every day, right from the apartment. But if we wanted to see them now, we'd have to climb.

"They look like stones," Bård warned us in his heavily inflected, breathy English, waving vaguely at the scree slopes. "If they are behind a bump, you won't see them, because they don't make a sound."

The next morning, we filled our water bottles, doused ourselves with DEET, and donned our new mesh head nets. I'd like to say we looked forward to our climb with barely contained excitement, but that would be far from the truth. The kids protested. Vociferously. We finally got out the door with a promise: gadget time when we get back.

By the time we reached the edge of the forest, a swirling mix of bloodsuckers was bombing us, giant horseflies with amber-striped abdomens, mosquitoes, blackflies, something else that's shaped like a small grey bullet, with a telltale spike below the eyes. This, I later learned from the internet, is known across Europe's northern latitudes as a cleg fly, from the Old Norse term for these pests, *kleggi*. They like to zip to a landing and then sit still, sharpening their blades before they bludgeon their way into our blood supply. They seemed unafraid of death.

After about a mile along a winding trail made by roaming sheep, it became clear that we weren't really hiking—the bugs were literally chasing us up the mountain. To spare the kids, I'd taken up position at the rear, where the insects drafted in our wake. I was wearing jeans, but the horseflies and cleg flies didn't care. They landed right on the back of my thighs, whisper soft, and I only noticed when I felt them augering through the fabric into my flesh.

"Man-o-man," I complained, my hand swishing across my backside like a makeshift tail. "This is the hike from hell."

Mistake. The kids picked that one up. *The hike from hell* they chanted from under the black billows of their head nets. Dad forced us to go on *the hike from hell. In the fly forest.*

The scary thing was, from a reindeer's perspective, this was no *rakka*. Reindeer actually migrate here in the summer to escape the explosion of bugs in the interior. The *fly forest?* More like a walk in the park.

We stopped to get our bearings by a small stream, swollen with snowmelt. Bunchberry, a miniature dogwood relative with greenish blooms, carpeted the ground. Birds flitted through the pale veneer of a birch canopy above us, fieldfares, like chunky, mottled versions of our robin, and sleek black redstarts.

The wind-blasted limbs of the birches were low enough that I could see the birds' beaks were hinged open—they were panting in the heat like us. We stared at the unmoving boulders up on the plateau for as long as we could stand it, which was probably less than a minute.

Back at the farm, the kids seized their electronic reward with eager delight.

~

The world's only research station devoted exclusively to the study of *Rangifer tarandi* lies in a remote corner of northern Finland, on a river bend just west of an enormous lake and the Finnish Sámi cultural center of Inari. To get there, we crossed the stark and empty Finnmark Plateau, a place even the hotel reservationist in the regional capital of Kautokeino, or Guovdageaidnu in the Sámi language, advised us to avoid at this time of year: the bugs were terrible, the reindeer were gone and the only herders around were those shackled with responsibilities.

The station, with its herd of 150 reindeer and 16 square miles of pasture, was managed by Mauri Niemenen, who had published on everything from contamination levels in lichen to the number of reindeer that die in traffic accidents each year. Niemenen, who is sturdily built, with a sun-bronzed dome of a forehead curving over a thick silver beard and bright blue eyes, offered me a chair in his study, a low-slung wooden cabin on the shore of a lake whose name in Sámi means "not too big, not too small." The hide of a brindled cow carpeted the floor in front of the wood stove, while a collection of pelts hung from the wall, the pale luxuriant fur of silver and arctic fox, and the creamy, grizzled fur of a much larger predator, a wolf. Next to the furs were several examples of the braided lassos reindeer herders from different regions employ during earmarking. One, from an indigenous community in Russia, was made of twisted reindeer leather, darkened with age and use.

Niemenen has been around long enough to see reindeer herding change over time, and he was worried by what he was seeing. For several years, researchers had been using satellite imagery to inventory winter pasture, and what they'd found in Scandinavia, he said, was evidence of severe overgrazing.

"In Scandinavia, the average quantity of available lichen is one hundred and twenty kilograms dry weight per hectare," he said earnestly, whisking away a stray mosquito with a meaty, weathered hand. "Cross the border into Russia and it's fifty thousand kilograms per hectare." His eyes widened for emphasis, making sure I'd understood his English. "Fifty *thousand* kilograms per hectare."

That amounted to a 99% reduction in a critical source of winter calories on this side of the border, leaving herders with a tough choice: either feed their reindeer expensive hay or pellets from more southerly climes, or watch them weaken and starve. In Finland one recent winter, all 56 reindeer cooperatives had to give their herds four months of supplemental feeding. Some of the animals were simply too remote to reach, even by helicopter.

Before there were codified national boundaries in this far northern region, reindeer herds moved across a vast wilderness. They could head south into the mountains of Sweden, north along the Norwegian coast, east into the dense forests of Finland and beyond into Siberia, following ancient migration routes that took no notice of political borders. Nomadic people could follow the herds, trading with sedentary communities along the coast without worrying about a passport.

That changed, of course, as regions evolved into separate nation states. The establishment of national borders not only altered reindeer migration patterns; it effectively divided reindeer husbandry along national lines, and now each country has its own reindeer management regime, and reindeer politics, too. Unlike Norway and Sweden, for example, in Finland reindeer ownership is not restricted to the Sámi, and 80% of the 4,000 herders in the country are not of Sámi origin. Finnish herding practices are fundamentally different too. Without access to summer pastures along Norway's coast, Finnish herders tend to keep their animals in a fixed location rather than migrating with the seasons, an approach some dismiss as farming rather than herding.

From Niemenen's perspective, the problem is that the deer are simply too numerous for the land to support. Although their policies vary to a degree, Scandinavian governments have generally encouraged reindeer herders to maintain larger populations in order to make a living exclusively from herding. To be eligible for subsidies in Finland, for example, you must own over 80 reindeer. As a result, average herd size has increased from 80 animals when Niemenen started studying the reindeer business 40 years ago, to 100 head per herder today. Four decades ago, there were 7,500 reindeer herders in Finland, but that number has declined by nearly half, even as the overall number of reindeer climbed.

It's a sobering reminder of the question of scale, and the ecological limits imposed by the very nature of the enterprise, especially during a time when northern latitudes are experiencing more extreme effects from climate change. There are probably some things you could do to expand reindeer production, such as expanding the winter range available to herders, and protecting and improving the pastures that already exist. Still, herding will never be a replacement for industrial agriculture. You can't scale up herding to feed a population of billions; it's a small, subsistence-based practice, producing food enough for a hardy few. Most of us urbanites will never taste reindeer purchased from the Whole Foods meat counter, let alone consume it regularly. Nor should we.

~

That line between farming and herding—I kept thinking about it after we stayed at Nord Hestnes Gard. Farming was once the cornerstone of colonization, which is perhaps why it is so controversial in the reindeer herding area, where the Sámi have only recently achieved some measure of autonomy. At the same time, farming demands more time with the animals, tending to their needs since you can't let them roam, strengthening the bonds between species that are the core of domestication. In that sense, it's closer to what reindeer herding was like in the not-too-distant past. One black and white photo taken early in the 20th century, for example, before the so-called "snowmobile revolution," shows a reindeer pack train waiting to move; the animals' flanks are loaded with hefty sacks of gear. Another shows a Sámi woman with pale eyes and a tight-fitting cap, offering a sardonic smile for the camera as she tilts her head and reaches under the udders of a tethered reindeer with a milk bowl. These nomadic people lived in close proximity to the animals they tended, and although their herds were smaller they relied upon them for more.

It was a reindeer farm that finally allowed us to see human and reindeer together. We arrived in the early afternoon heat, deep in the woods near the end of a dirt road. One of Niemenen's neighbors, Tuula Airamo, a member of the reindeer collective in Ivalo, greeted us at the steps to the small clapboard house she inherited from her parents. She was tall and thin, with prominent cheekbones and twinkling eyes, her shoulder-length hair tucked up high in a clip. There were no reindeer in sight.

I flung open the car door, ready to sprint for the porch. But the kids kept their seat belts fastened.

"We can see from here," Brook said uneasily.

It was my fault. Another reckless binge of parental paranoia. Do you know what a warble is? Not the song produced by the family of small insectivorous birds—I mean the skin affliction. Having studied a bit of medieval lit, I was familiar with the *whelkes* and *knobbes*, *carbuncles* and *buboes* that could pop up on somebody's face, but the warble sounded far worse. It's a boil, essentially, a painful lump that appears on the face or the neck. Which is gruesome enough, but this boil moves; it migrates, erupting on the jawline, the cheek, the temple, getting ever closer to the eye socket. Inside is a maggot, exuding enzymes that dissolve the connective tissues of the skin, feeding on your flesh as it tunnels along, looking for room to stretch out and grow. Which it typically finds at the back of your eye, where the retina meets the optic nerve. Extracting these larvae can cause blindness in the afflicted eye.

How do you contract this gruesome disease? Looking at reindeer on a hot summer day. The reindeer warble fly, *Hypoderma tarandi*, looks like a bumblebee—it's furry and striped and kind of portly. But this insect isn't looking for flowers. Its life cycle depends on a reindeer host; the larvae emerge from eggs laid in the animal's fur, drill into the flesh and feed until the following spring, when they emerge again to pupate, hatch and afflict another host.

Warble flies are most active on warm sunny days, when they may buzz around for miles in search of their migratory quarry. However, once they find what they're looking for, they tend to stick around. They can be found in especially high concentrations where reindeer are confined, as they sometimes are near tourist outposts. Penned reindeer often get extremely agitated when these bugs are flying around, rearing up, kicking, and thrashing to keep them away.

As a tourist, you wouldn't know why these bumblebees were looping around your head, and with all the black flies and horse flies and cleg flies and mosquitoes buzzing around too, you might not even notice them, even when they landed. Just a moment's pause, a little exploratory crawl, a bit of ovipository labor, and that's it, you're an accidental host.

This, according to a report in the Swedish science journal *Läkartidningen*, is what happened to one ten-year-old boy who visited a reindeer pen while on a summer hiking vacation with his family. He saw the reindeer freaking out but didn't know why. Didn't know the flies had been landing in his hair until he developed big red bumps that seemed to move along the hairline on his forehead. The accompanying photo showed a profile of a boy about Brook's age, with similarly fair hair and what looked like the kind of lump kids often get from banging their head.

"This case description shows that even a short-term stay in the reindeer herding areas can lead to infestation with the larvae where many flies are present," the authors concluded. "Something that could be a bigger health problem than is commonly known." Indeed, after reading this report, a Swedish pediatrician diagnosed several more cases in his young patients that otherwise might have gone unrecognized.

Ahh, the wonders of the internet. Back at the farm in Norway, I had been sitting at the table in our apartment, ostensibly trying to find out more about the life history of cleg flies, when as is so often the case online, I stumbled across something far worse. As I was contemplating the horror of a medical photo that showed a glistening, lanky larvae squeezed between the tips of a pair of tweezers with what appeared to be the halo of someone's illuminated eyeball in the background, Brook appeared quietly at my shoulder. I didn't know he was there at first, and then it took me a moment to disengage and realize he shouldn't be looking at this kind of thing.

He pretended he hadn't seen it, but later he piped up from the backseat.

"What was that you were reading about a fly?"

I told him, as convincingly as I could, that this kind of insect attack is rare, so rare that doctors here don't even recognize it. I almost believed it myself. I mean, what were the odds, really? Except weren't we a family of tourists, with kids the same age as in the photos? Wasn't it within the realm of possibility that we would try to get a closer look at some reindeer on a warm sunny day? I'd found something to fret about. It was almost comforting.

Now the kids were glued to their car seats with no sign of budging.

"There's no warble flies out there," I told them, using my authoritative father-knows-best voice. "It's too warm for them. But you can wear the head nets if you want. They can't get at your scalp then. All right?"

They looked at each other. A mosquito was already pinging at the glass of Brook's window.

"Coats too?"

Fine, I said, glad I didn't have to play my trump card: no reindeer, no iPad.

There was a small lake visible at the back of the property, and its rippling, slightly tannic surface reminded me of Northern Maine. Which said one thing to me: bugs, lots and lots of bugs. I saw Airamo had some cans of repellent lined up on the porch, but she wasn't bothered—the sleeves of her flannel shirt were rolled up and the neck was open, exposing a triangle of tanned flesh.

"You can take your coats off," she told the kids, who had donned raincoats despite the heat of the midday sun. "Layers don't help. You just tuck in your shirt and your pant legs..."

They weren't buying it.

Araimo led us behind the house to a spacious but empty pasture. At the edge of the trees sat a small outbuilding with roughhewn, split rail walls and a low, galvanized roof. It looked a lot like the simple, rustic structures in the outdoor exhibit at the Sámi museum we'd visited, like the turf-roofed, cramped barns where the Sámi kept their herd when they weren't on the move. Her father, Airamo said, was Coastal Sámi, one of the branches of the indigenous community who relied on the sea rather than the interior.

She pushed aside a blanket and motioned us inside. Light poured in through the chinks between the rails, and as our eyes adjusted we saw 12 antlered heads nodding in the hazy half-light. They snuffled as they pushed their blunt equine muzzles toward us with wide staring eyes, dodging each others' antlers, waiting for us to hand out some of the fireweed and grass Airamo has collected in a wheelbarrow by the door. I expected a wave of barnyard smells to wash over us, something sour and grassy like penned sheep or the acrid odor of goats, or the musty sweetness of the horse stalls back at Nord Hesnes Gard. But there seemed to be nothing in the air here but dust and wisps of winter hair. Even the bugs were absent; only a handful of mosquitoes seemed willing to venture into the shady interior.

The biggest male was 12 years old; he had extra turrets protruding from his velvet antlers, which in these tight quarters meant he was constantly getting nudged aside with the hollow, velvet-softened clap of bone on bone. Antlers are the fastest growing mammalian tissue, and over the course of the short Arctic summer, they practically erupt in a hormonally charged, capillary-coated reef of bone. In the biggest bulls, they can reach four feet in length and weigh nearly 30 pounds each, making them the largest antlers, relative to body size, of all deer species. As their testosterone levels peak in the fall, males smash each other with these weapons, jousting for the right to mate with females, and then cast them aside as winter arrives.

Calves are born in the spring, like the three in Araimo's herd: two brown and wobbly and just a month old, and another as pale as butter, bigger and bolder at three months. Snowman was his name; he and Snow White, his mother, were new additions to the herd.

Araimo let Snowman out with us, and he nudged between my knees like a baby goat while the kids tentatively stroked his back. His fur was slightly curly and dense as a layer of felt, nothing like the thick, almost Styrofoam-like strands of an adult whitetail deer's coat. Snow White didn't seem to mind the separation. Airamo pointed out the red thread dangling from Snow White's ear, which signified her transfer from one herd to another. Each reindeer owner has a different earmark, a series of notches cut into the upper and lower curves of the calves' ear in the late spring, at one of the big communal events of the year. Airamo's earmark was a series of three, then two notches above and below, a scalloping most visible in the ear of the big male.

We took turns poking the fireweed through the slats in the fence. The stems felt like a rope sliding taut against the palm; kids and calves engaged in a tug-of-war between the slats, and when it was stripped clean, they both let go. The big male antlered his way in, snuffled the fireweed and declined it. She patted his neck as he snuffled at her palm. He was clearly her favorite.

"Each year we have to choose two or three to slaughter," she said, with a touch of melancholy in her voice that made it seem as if it might be the big male's turn this year. "It's very difficult, but it's the rules of the cooperative. We use a stun gun—it's not traditional, but they don't suffer. I say they have a good life here, and when the end comes, it's quick and they don't know."

I thought of what it must be like to have steaks from an animal like this male in your freezer. I thought of Beatrice, feeding the calf, not wanting to entertain the idea of eating this meat. What is it that we're trying to model with biomimetic meat? It's this moment with the animal, and the product in the cooler. But nothing of the process in between. We embrace the practice of nurturing and deny the practice of slaughtering. We want to disconnect the animal from its body.

Araimo lead us back outside. She wanted to show us how tame the animals were, how obedient. They would come when she called their names.

She picked up a small silver bell and began tolling their names in a singsong voice, but the jingle only seemed to summon the horseflies and cleg flies, who were on us immediately. Or were they warble flies?

"We only get a few of those," Airamo assured us. "We can give injections, but that's costly. Most years we just pick one or two off in the spring." She made a plucking movement with her fingers, as if extracting a bit of lint.

More clapping, more calling of names in melodious Finnish, but no movement from the shed.

"They don't want go out now," she said, a little flustered. "They stay inside away from the bugs."

Forget the bell. Airamo had to resort to "reindeer candy." She filled a bowl with pellets and rattled them around, calling. Out trotted the big male, followed closely by several yearlings, who head briskly for the trough. The big male paused, sensing he had room, and reached up with a rear hoof, contorting himself in a three-legged backward bow as he slid a fat gland along the big curves of his antlers, greasing the velvet. The bugs, meanwhile, were going crazy. Big horseflies were bombing us as we talked, the blackflies were simmering, we were twitching and swatting like crazy people.

It might even have qualified as a *rakka*.

Airamo wanted to show us a progression of antlers she's saved, from a yearling's spikes to the broad spread of a mature bull, all the way to the shriveled tines of an elderly bull in decline, but we couldn't pause for more than a glance at the display she'd laid out. The yearlings were dashing back inside the shed; our kids were sprinting for the car. As a keepsake, she gave us two small bits of antler, bone you can harvest without harming the animal.

~

About 50 miles down the road to Kilpisjarvi, just past a cluster of boarded up cabins that passed for a town in those parts, the highway turned to dirt. We were still 40 miles of birch forest, marsh, and tundra from the junction with another road, 40 miles of smooth clay punctuated with brain-rattling washboards. No traffic in either direction. Nobody out there in the Finnish wilderness but us. The kids responded to the solitude and scenic monotony by clamoring for the gadgets.

Suddenly, up ahead, reindeer were everywhere. The road ahead was blocked by hundreds of cows and calves, trotting forward with their loose-hipped, almost stumbling gait, a motion due in part to the physiognomy of these animals' hooves. Reindeer, more clearly than other ungulates, have feet. Big, splayed feet, with prominent toes. The two big cloves of each hoof are thatched with bristles, and their enlarged dew claws, which sit above and to the rear of the main hooves, spread out to provide lateral stability in the snow. The hoof pads harden along the edges in the winter so they stay upright on ice, like crampon-equipped Sorels, then soften in the summer so they can cross the damp sponge of the tundra.

Maybe a hundred more cows and calves were grazing further up the slope as it stretched away in undulating folds of willow scrub and tundra to the scree of the ridge. It looked like a wild scene. There was no fence, no barbed wire, but some of the adults had bells around their necks, so the herd sounded like a deep-toned wind chime clanking atmospherically in the breeze.

Rangifer are the only domesticated species of deer, the sole species tractable enough to be harnessed and milked. The reason, experts believe, lies in their behavior. Unlike other cervid species, reindeer aren't programmed to startle and flee when they detect a predator since there's no place to hide out on the tundra. Instead, they gather themselves into a dense herd and follow the leader, a behavior pattern that made them amenable to the guidance of humans.

The herd surrounded and engulfed the car, like an amoeba ingesting some tidbit, and we seemed to float along inside. The calves grunted when they wandered away from their mothers. They wriggled their tails just like lambs when they nudged their mother's teats, nursing in front of the car. No herders in sight: if it wasn't for the scattered clank of the bells, you would've thought we were in Alaska.

The kids were indifferent to the magic, the serendipity, the thrill in my voice. They were playing *Minecraft*.

"Guys," I entreated them. "Look! Look out the window!"

They nodded that familiar, preoccupied, half-listening nod. *Uh-huh. Whatever.* I had to restrain myself from acting out a fantasy that involved flinging that glass and titanium wonder into the nearest willow thicket, to be excavated centuries hence by some archaeologist who would marvel at the extent of our habitual screen time. *Even here, in the midst of what would have been a wilderness, they still carried their screens…*

"Off," I growled. "One, two, three…"

~

We're going to head in the other direction now, out of the wilderness, into the realm of cellular agriculture. Let's start by unburdening ourselves of a simple but weighty assumption: meat comes from the body of an animal that has been born and raised and slaughtered, then butchered before it arrives on our plate. Instead, we might ask, "What is meat, actually?"

There's a competition happening right now to create the most convincing biomimetic model of meat. It's a competition between performances of burgeriness, resonant with all the ontological and epistemological and ethical complications that surround the production and consumption of animal flesh. The biomimetic process behind these performances should sound familiar—just as the gait of a dog or the swarming of ants can be represented in robotic form by first developing a detailed model of the source, then translating that model into a different medium, so the physical and chemical properties of a burger can be modeled and mirrored. Of the two competing paradigms, the one known as "cultured meat" got some early press but hasn't gained traction in the marketplace yet, while the other, "molecular meat," is already out there in fast food joints and coming soon to grocery stores.

Cultured meat is made up of stem cells taken from a donor animal and grown in a "bioreactor" that simulates the chemical and physical conditions of the original tissue in the body. These cells can then be harvested and consumed without ever harming the original donor—in theory, cells could be taken from a small sample from a single animal and grown in vast quantities. In essence this approach recreates the animal body outside the body, minus any cognitive or sensory apparatus. In a striking re-enactment of Descartes disarticulation of brain and body, it's strictly the cells of the body that are reproduced here, not those in the brain or the olfactory bulb. There's no sensing, thinking, or feeling involved.

I attended the first international symposium on cultured meat, which took place on the campus of Maastricht University, about two hours south of Amsterdam, and what was clear from the proceedings was that there was a lot of commercial interest, a lot of technical progress, but also a number of significant challenges. Again, it was mostly a question of scale: how can you create a growth medium full of the hormones and nutrients these cells need to proliferate rapidly, that's also cost-effective and healthy enough for humans to consume? Then there's the question of taste and texture. As Mark Post, cultured meat's most notable figure, said in a lecture, hamburger is just the beginning. You're trying to mimic a mosaic of different types of cells, he said. He dreams of culturing steak, that edible conglomerate of fat and muscle fibers threaded by blood vessels, nerves and connective tissue. It remains a lofty goal; few people, to date, have ever taken a bite of cultured meat in any form.

The other model, molecular meat, mimics the properties of animal cells rather than trying to grow them. Cells are made up of molecules with various properties, some slippery, some stretchy, some tangy or bitter. If you could inventory the chemical compounds that make a burger feel and taste like a burger and find alternative sources for them, then you might be able to mirror the original without involving any animal cells.

That's the premise behind Impossible Foods, the biotech start-up that has already made waves in the food industry with their vegan burger that bleeds—you can try one at Burger King these days, and ads featuring the testimonials of elated carnivores make frequent appearances on all sorts of platforms. The special sauce is not beet juice or some other red food coloring, but actual blood. Vegan blood, no less, made from analogues of the oxygen-bearing proteins that flow through our veins, but with no bovine cells or bodies required. Let that settle in for a moment, through all the layers of culinary and anatomical expectation.

Unlike many start-ups, the company is noted for its transparency, and I flew out to the Bay Area to attend one of the events they host regularly for the press. To get to the company's industrial location, I followed the highway south of the airport behind a line of dump trucks loaded with construction waste. I was struck by how distant this landscape all felt from the shadowy interior of Tuula Araimo's barn. No green pastures flecked with grazing livestock here; these were brownfields, backfilled marshes where pesticide and dynamite factories once stood. It's not just that no animals are required to produce molecular meat; you don't need the conventional agricultural landscape either. Which, rationally speaking, is great news. We already devote an outsized proportion of the planet's resources to raising cattle, and the global appetite for meat is growing by leaps and bounds as living standards and human population continue to rise. What's left for us to slash and burn, besides rainforest and derelict strip malls?

And yet, call it sentimental or ecomimetic, but I could feel myself resisting this anti-pastoral scenery. Rationally, I know most livestock don't roam like reindeer. I've driven past the feedlots on the highway between Los Angeles and San Francisco; I know that's the reality. But the further we go down this technological path, it seems, the more "natural" my cravings become. The question is whether metaphor can or should assuage that hunger.

Outside the company's front door, a media frenzy was set to unfold. A camera crew from Japan sat nursing their jet lag next to a crew from Korea, while a production team from a German barbecue program gathered footage of the host, a bald giant of a man in a loose black t-shirt, strolling casually toward the entrance.

Eventually, we got the signal—they were ready for us. We trooped through a cavernous open hangar of a room where about a hundred employees sat at long rows of computer monitors and squeezed into a small meeting room. Inside, our hosts had laid out the makings of a food demo on a narrow buffet table—an electric griddle, glass mixing bowl, some wooden utensils and an array of glass ramekins filled with various liquids, pastes and flakes arranged on squares of polished wood. Each ingredient was labelled in white on a small piece of slate, like miniature versions of a blackboard in a restaurant where the menu changes daily according to what's fresh and in season.

"This is an all-natural process," Celeste Holz-Schietinger, a principal scientist who has been with the company since it began, assured us. "We have five main ingredients, all found in your grocery store, except one key ingredient." She was dressed in black jeans and a heathered grey scoop-neck t-shirt rather than laboratory garb, in keeping with the wood, glass and slate. Natural. That was the mantra here. No petri dish, no lab bench. This could be your kitchen countertop. Which was interesting, when you consider that a hot trend in the culinary trade not so long ago was molecular gastronomy, whose working set of principles seemed to involve bringing the tools and sensations of the lab into the dining room. In these establishments, the diner might've been invited to sample synthetic champagne squirted into a Fisherbrand 250 ml beaker from a hefty syringe, or eat a picture of a cow, printed on edible paper and flavored like steak. Molecular was postmodernist cool, although it was never intended for the masses.

Holz-Schietinger described the guiding principle of molecular meat as if taking a page from a phenomenology tome—beef, she said, is an *experience,* brought about by a series of physical and chemical transitions. A biomimetic phenomenon, in other words, not bioinspired. No need to make allowances for whatever might be lost in translation.

First, she said, we experience the texture, which goes from soft and malleable to firm. Fat sizzles as it melts into the juiciness we expect, while the flavor profile changes from metallic to caramelized as the color transitions from red to brown. The research team looked at each of these transitions at the molecular level, then searched for analogues that didn't come from an animal. That, she said, is what distinguishes this product from the veggie burgers already stocking the freezer shelves—those patties don't mimic the transitions that make burgers taste like burgers.

"Our Impossible Burger is plant-based *meat,*" she affirmed, as she took us through the ingredients on display. "The exact thing that causes the flavor reaction in a cow beef, it's the exact same transition. We're not trying to add in some flavors to try and match beef. We actually have the exact same reactions to the exact same array of flavor compounds."

This was the strictest possible approach to mimesis, in other words. Not bioinspired. Not *as if,* but *is.* The goal was not dictated by the biology per se; the purpose was not to learn more about cow physiology at the molecular level, but to satisfy the pre-existing expectations we have for the experience of eating a hamburger.

The host of the German barbecue show raised his hand.

"You just killed the cowboy, right?"

Witty, but also a bit edgy. The cameras were rolling. He was alert to what she'd left out of this performance: the practices of meat production, not just the animal itself. The phenomenon had been disarticulated from the practices that normally produce it, but it was nevertheless an embodied metaphor.

"The cowboy can enjoy the cow," Holz-Schietinger replied unflappably, as she poured some flakes that looked like *Cream of Wheat* into the mixing bowl, stirring with the spatula as she talked. These were binding and texture agents that absorb water and mimic the "chew down" properties of muscle tissue. Then she picked up the ramekin that looked like it was brimming with wine, the one marked *leghemoglobin*.

This was the one ingredient you can't get in the local supermarket. The story of leghemoglobin begins with the company's founder, Stanford biochemistry professor Pat Brown, worrying about the global ecological crisis, about climate change and the outsized impact of conventional meat production, when he had something of an epiphany: the missing ingredient in your usual veggie burger, the one thing that any carnivore's taste buds were sure to miss, was the flavor of blood. Meat tastes like meat because of *heme*, the iron-rich molecule which, when it forms part of the protein chain known as hemoglobin, turns red blood cells red and transports oxygen and carbon dioxide throughout the body. Heme can be found, in varying amounts, in virtually all organisms—sperm whales, I was told, which dive deep and stay there as they hunt for squid, have one of the highest percentages of heme in their flesh. But as Brown discovered, trace amounts of heme can even be found in plants. His research team identified minute quantities of the red stuff lurking in the root system of soybeans, and from there the question became one of scale. How can you get enough of this vegan alternative to transform the meat industry?

Holz-Schietinger splashed heme into the concoction and stirred—there was an element of horror movie here as the pancake batter turned a familiar red, a bit of *Carrie*'s bucket of blood, although the liquid gleamed like juice, the blood of grapes, not the gobbets of semi-coagulated stuff you might see the cleanup crew sweeping up with a broom after a bullfight. Blood is full of other compounds, like the platelets that clump to staunch a wound, but what we were dealing with here was a purified version, the essence of a bloody flavor.

She added the oily looking liquid next, explaining that heme needs to react with various amino acids and sugars to form the compounds you'd find in beef, then a carbohydrate gelling agent derived from the tubers of a tropical plant, and finally the "plant adipose tissue:" a generous smear of congealed coconut oil, which acts as a "flavor delivery system" by trapping aromas inside the burger until we release them with our chomping teeth.

With each addition, what was in that bowl looked more and more the part—it had the pinkish color of ground muscle tissue, studded with pearly tidbits of what looked like fat, or whatever those little white nuggets are in actual burger meat—I've never been inclined to look all that closely, to be honest. This vegan meat stuck to the spoon in the mushy but also slightly clingy way of ground up raw flesh.

The German host raised his hand.

"Can we taste?"

I sensed a titillating ripple of aversion go through the crowd as we realized what he meant, as if we'd strayed into taboo territory. I felt it too, as if a cautionary hand now rested on my shoulder. *Raw? Raw hamburger? Are you crazy?*

"Yes, you certainly can," Holz-Schietinger replied. "Several of our chefs are experimenting with serving this tartare." There's no slaughterhouse involved, she reminded us, no contamination with animal waste, no devastating, antibiotic resistant E. coli strain hitching a ride into our microbiome. You could eat this hamburger rare or well done. Right out of the bowl.

The TV host reached into the vegan flesh with a tiny silver spoon, hoisted out a good-sized pink lump, and after a brief dramatic pause for the cameras, inserted it into his mouth. Soon he was nodding appreciatively and smacking his lips.

"Unbelievable!"

I was next.

"The heme molecule is identical to what's in a cow," Holz-Schietinger reiterated as I dug in with my spoon, a claim whose sensory validity I can confirm even if, in this moment of gastronomic truth, that's exactly what I was trying mightily to erase from my mind. She kept invoking the cow, when normally, in my mind, I don't eat cows. I eat burgers. Yet at the same time, I knew this was just a particularly potent biomimetic metaphor—it wasn't actual cow flesh.

Not cow, I repeated to myself silently, and without much effect, as the spoon approached my nose. *Potatoes and coconut. Potatoes and coconut.* But try as I might, I couldn't dispel the notion that this was raw beef. It smelled like raw beef, like warm pennies in a sweaty palm. It squished like raw beef. It was red and glistening, just like the ground chuck that sits in mounds in the meat counter case.

It tasted just like you might imagine a spoon of raw hamburger to taste, which is to say, meaty. As in muscle, fat and blood, not potatoes and coconut. It's psychosomatic, I know, but I couldn't help imagining myself at a supermarket meat counter, waving away the wax paper wrapper with the safe handling instructions stuck to it. *That's okay, you don't have to wrap it; I'm just going to eat it right now...*

The queasiness wasn't helped when the TV host asked if we could sample the blood too.

Holz-Schietinger readily agreed, although she warned us: it's potent.

Not cow, I repeated, when my turn came. *Not cow.*

What does a spoonful of blood taste like? Umami, maybe? Again, I'm at a loss for a comparison. When was the last time I took a sip? The only possible likeness

I can think of is the inadvertent taste of my own accidentally spilled platelets, but then again, when was the last time I had a bloody nose and thought, *hmm… that's an interesting flavor?*

Okay, if umami it is, then vegan blood is umami to the nth power. It tasted briny and metallic and mouth puckeringly gamy; it tasted like what I imagine it would be like to suck on a beef bouillon cube or slurp the decanted broth from the vats of fermenting fish that eventually decompose into condiments like fish sauce.

After the electric nonstick griddle heated up, Holz-Schietinger held aloft a stack of three-ounce patties and scooped up the top one with her spatula.

Down it went with a faint sizzle.

"You can probably start to smell some of the aromas, people in the front rows…"

We could. It smelled like someone was cooking a hamburger. More specifically, all the chemical compounds being generated as the heme molecule transitioned from red to brown were billowing over us in the steam. It smelled good, and it was a relief to have my headspace refreshed with a familiar sensory experience that conjured backyard grills, not barnyard cattle.

I couldn't get too comfortable, however, because Holz-Schietinger kept reminding us that the heme in the dish was the same as that in the vessels of the cow, a point that prompted a question from the German host.

"You call it beef?"

"Yes, we call it beef."

"Because?"

"Because why define beef by what it comes from instead of the sensory experience?"

A loaded question, and one that gets to the heart of this journey. Why limit the definition of beef to flesh harvested from a cow—to the animal source, in other words—instead of embracing biomimetic sources whose experience mirrors beef? Or, to put it another way, is beef strictly a product we consume, or is it also a set of practices through which animals convert vegetation into bodies, which we then convert into the patty on the plate?

Impossible Foods makes no bones about its mission, which is to replace cattle as the world's source of beef by 2035 and eliminate the burden that conventional beef production places on the planet. Ambitious is perhaps too mild a word for the exponential growth that goal will require, especially since global meat consumption is projected to more than double by 2050. But the ecological stakes are high. Conventional animal farms already use one-third of the ice-free land on the planet, and ruminating cows are notoriously prodigious producers of greenhouse gases, particularly clouds of heat-trapping methane that erupt from both ends. Raising cattle is responsible for between 14.5% and 18% of the greenhouse gases produced by human activity each year, and we haven't even mentioned the water required or the rainforests razed each year for pasture and feed crops.

According to an independent report commissioned by the company, the 2.0 version of the Impossible Burger, which uses soy protein instead of wheat, requires 87% less water and 96% less land than a burger made from cows; it also results in 89% fewer greenhouse gas emissions. This dramatically smaller ecological footprint comes about primarily through the efficiencies gained by removing cows from the production equation; cows did not evolve to be efficient producers of flesh for our consumption. The efficiencies gained in feed and water requirements, as well as reductions in emissions, more than offset the increases in energy usage associated with producing vegan meat at industrial scale.

The Impossible Burger is not a complete overhaul of our food delivery system. It's not slow food. People are still pulling up to the drive-thru to get their meal. The fundamental premise of this burger is that the experience stays the same—not just the taste, but the familiar ritual of getting off at the exit and grabbing a bite to eat. It's not meant to entice vegans into enjoying fast food; it's meant to reach carnivores who are already pulling into the lot.

Given the potential ecological benefits, I'd be perfectly happy knowing that all the meat in fast food burgers, tacos and pizza toppings was actually vegan. But does that inevitably mean killing the cowboy and the reindeer herder? If scale is no longer the goal for livestock production, then our relationship with the species we raise and eat will evolve into something different. Domestication itself isn't static; it's a set of living practices. The Dutch philosopher Cor van der Weele was among the first to imagine what the future of our relationship with livestock might look like. She also spoke at the Maastricht symposium, and over a carnivore-friendly banquet held in an atmospheric cave outside the city, we talked at length about a vision she has nicknamed "the pig in the backyard." What van der Weele discovered, through a series of visioning workshops she'd organized, was that potential consumers kept coming back to the idea that having a pig rooting around out back was a traditional fixture of Dutch village life, one they were loath to lose.

Van der Weele envisioned a different possibility: scrape a few donor cells from the hide of a happy neighborhood hog, pop them in a countertop bioreactor, and let them grow into the pork you'll cook with later, something akin to a crock pot, or a jar of sourdough starter. "The degree of shared enthusiasm in response to this idea was remarkable," van der Weele wrote of the workshop. "It was so large that the preferred future of cultured meat was completely clear, as far as the participants of this workshop were concerned. A combination of joy, inspiration and amazement characterized the atmosphere." Of course, none of the workshop participants had ever tasted this rarified flesh, and the countertop bioreactor remains a concept whose realization may still be a long way off, but nevertheless, *joy, inspiration* and *amazement*? I've heard foodies use terms like these to gush about an exquisitely prepared steak—about the product, in other words. But I've never heard anyone describe the practice of meat production in this way.

~

After lunching on burgers, we donned white lab coats and plastic glasses and followed the company's R&D director, Chris Davis, into the research lab, an expansive space about the size of a basketball court, permeated by the hum of ventilators and the faintly acrid smell of acetone wash.

"Why is meat so delicious?" Davis inquired rhetorically, as we passed a technician in a hairnet hunched over a laptop and an array of polaroids, each a portrait of a patty with a slightly different color tone due to gradations in the fat content. We paused for a brief presentation in front of a gas chromatography-mass spectrometry unit, which can be used to identify the chemical compounds given off by a particular sample. It was attached to a "headspace" simulator, a glass booth which seals off and concentrates volatile organic compounds in the air, then funnels a sample to a "nose port." Davis said they had found several hundred compounds given off by ground up cow flesh, but ironically none of them alone smell of beef: they'd identified barnyard smells, the sweet ferment of different ripe fruits and the fusty odor of mushrooms, but all these chemical nuances are assembled in our olfactory systems into the whole picture we know as the smell of beef.

Impossible Foods' approach to beef—breaking it down to the chemical level, identifying the key compounds involved in the flavor and then synthesizing those compounds from environmentally friendly, "natural" sources—has a long and commercially successful history. There's an entire industry devoted to identifying the flavor compounds in food and then finding ways to produce the same compounds in cheaper and scalable quantities. Depending on the source, these flavors can be labeled natural or artificial; the key phenol compound in vanilla, for example, can be extracted from the oily brown pods of a tropical orchid, at great expense, or synthesized far more cheaply from compounds found in wood pulp and petroleum derivatives.

In this case, ironically, going the natural flavoring route means getting into the livestock business—microbial livestock of the *Komagataella* yeast variety to be more precise, which must be tended and fed before they can be harvested. Vegan meat may be an experience, but like cow-based beef it relies on a set of agricultural practices. The scenery is different, the husbandry techniques are different, the organism whose microfiltered body products we consume is different, but it's still agriculture. Cellular agriculture.

How Impossible Foods' strain of *Komagataella pastoris* came to be sanguineous is the high-tech twist to Impossible Foods' method. In theory, you could harvest the ruddy soybean roots Pat Brown's team discovered growing in a field, extract their minute quantities of heme proteins, and eventually you'd have the same vermillion flavor compound I tasted earlier. But it wouldn't be worth it, ecologically or monetarily—the concentration of heme in these plants is so minute that you'd need to water and fertilize acres and acres of soybeans to extract enough of the protein. Like producing reindeer protein from lichen, it's a question of ecological constraints, and scale.

The way around this problem? Genetic modification. Not something your Victorian brewer or baker might do, but it's a fairly straightforward procedure in the 21st century. Take a snippet of genetic code from the DNA of soybeans and insert it into the DNA of a yeast species that's known to be amenable to propagation in research settings. Technically speaking, leghemoglobin is a complex protein, not a genetically modified organism, but its origins are nevertheless a contentious point, since so much environmental consciousness raising has already gone into branding "Frankenfoods" as risky to human and ecological health. You'll find "non-GMO" stamped prominently on the packaging of Beyond Meat burgers, for example, one of the Impossible Burger's chief competitors.

By making genetic modification the agent of ecological change, Impossible Foods is challenging this environmental orthodoxy. It's asking us to think differently about what we take to be "natural," to be sophisticated enough to make distinctions between different kinds of technology, between a genetically modified corn cultivar that allows fields to be drenched in herbicide and a genetically modified yeast that keeps carbon out of the atmosphere. It summons the ethical spirit of Janine Benyus's definition of biomimicry, the notion that by mimicking Nature, we discover environmentally sustainable solutions to vexing problems. It's why Holz-Schietinger's demo talk was liberally sprinkled with references to natural, simple, and familiar ingredients—the product itself, like most of the examples of biomimetics we've seen in this book, is difficult to categorize according to conventional definitions of the natural and unnatural. It's both.

I'd like to believe that the existence of alternative sources of beef will nudge us toward a more honest reckoning with where our food comes from—I like technology that makes us think more carefully about our relationship to the natural world, that's confrontational when it should be, and uncanny enough to raise questions. That's what biomimetic technology should do, when it's done well, and done for the right reasons.

~

After spending some time in the test kitchen watching technicians make patties, our group headed across the compound to the pilot plant, where the air smelled strongly of fermentation, like a pub on a Sunday morning. In one room were big stainless steel tanks where the yeast grows in a sweet fizzy mix of food grade sugar and oxygen. Once the colonies reach a certain concentration, the solution is sucked into a set of giant blenders, which "lyse" (a scientific term for the button that would be marked obliterate, if you're a SpongeBob fan) the yeast cells into a molecular slurry. The solution then heads to a plus-sized centrifuge to spin out most of the solids, and from there to a membrane-based filtration unit. The vertical tubes were clear, and we could see they were crimson inside, like giant arteries.

Standing in this plant was the technological equivalent, you might say, of ducking inside Tuula Araimo's barn, the moment in which you meet the prospective source of your food. For a bookish denizen of the great outdoors like me, it all felt a bit Willy Wonkaish—it wasn't the sensory experience of antlers and hooves in stripes of shadowy light, wasn't the chicken droppings whitewashing the steps at Nordre Hestnes Gård. It's not a replacement for that kind of meat, for the experience of raising it, for the long cultural history that accompanies a meal made with the flesh of animals raised in this way. There was no pig in the backyard. And yet, the tubes and tanks felt more in keeping with the contemporary way we eat.

Most of us eat packaged and plated artifacts—or at least we like to pretend that's all we're eating. This was a facility that grows artifacts. There was a logic there. There was no reason to avert my eyes. No need to pretend that the sentient animal is really just an eating machine, no need to hide what happens to animals as they're processed into food items.

At the same time, how different it all felt from our previous lab visits, all of which seemed to be asking us to imagine the outside world as we stood in the lab—imagine this ambulatory robot romping down the sun-dappled trail one day, kicking up a mechanical binkie, even if right now it just looks like a bundle of circuits and wires crossing a patch of linoleum. *We're going to make this cybernetic organism even more like the fauna you find outside,* the unspoken message seemed to be, *more convincing with each generation, more like an animal in every way.* What was happening here, however, was the disavowal of the slaughterhouse; the very fact that we were inside, witnessing the equivalent of the killing floor, was an indication that this place was not supposed to be anything like a conventional meat production plant. There was no attempt to reproduce the cattle car unloading its cargo; no convincing biomimetic cattle bellowing in fear. We were supposed to note the absence of animals here. And then sense their metaphoric presence, as a source of inspiration, back in the test kitchen.

That's how biomimetics works. It doesn't resolve all our cares; it makes what we take as given, as *natural,* feel mutable instead, subject to revision. The impossible, made possible. That's what came to mind when I took my first bite of the Impossible Burger: *this is what biomimetics tastes like.*

The outside of that vegan patty was bronzed by the grill and taut against the teeth, the inside moist and crumbly and suffused with the aromas of burgeriness. But it was the middle, where my teeth had left their mark, that matters. When culinary types talk about this meat, that's what they rave about, that color, how it registers as the real thing. *Beef,* that ruddiness seems to say; not necessarily *cow.* The things that were missing were things I don't tend to miss: no nuggets of gristle to pluck off the tongue, no lingering sense that somewhere back in the supply chain are faces whose eyes I'd rather not meet. At the same time, it conjured the sense of the red material at the meat counter, the artifact we know.

It was a burger. If you're thinking strictly in terms of the sensory experience, that pretty much says it all.

~

When through hikers reach the end of the Appalachian trail, they traditionally deposit a small token of their journey on the top of Mount Katahdin: a small stone, picked up at the start in Georgia, some 2,000 miles and many months behind them. It's a reminder of how far they've come, and an invitation to reflect on the meaning of the journey.

A biomimetic nature walk leaves us with something less tangible. There's no summit on this trail, no peak to bag, and yet we've reached a kind of vista, a point where it's customary to reflect. What emerges from an unconventional study of technology and nature? The creatures we've encountered are liminal beings, and we don't walk on trails that are separate from them—we too, this biomimetic praxis teaches us, are liminal beings. We belong most fully, perhaps, not in some untrammeled empty corner of the planet but in a landscape like this, not quite pastoral, not quite wilderness, not strictly technological, not entirely biotic. If there is a landscape where humans perform wildness, this is it, and it arises in these gestures of translation and representation that take us out of ourselves and make us self-aware at the same time. The lessons for how we should inhabit the Anthropocene come from navigating this uneasy terrain.

From where we are, you can see there's more territory ahead, faint in the distance like daubs of watercolor. There's more leaping and crawling and fluttering in those far fields, enough to fill new editions of a field guide to technological creatures, to add more entries to a technological life list.

The trail keeps going.

NOTES

Introduction: the biomimetic trail

2 **The term biomimetics**: Jon M. Harkness, "A Lifetime of Connections—Otto Herbert Schmitt, 1913–1998," *Physical Perspectives* 4 (2002): 456.

3 **In a review:** Julian F. V. Vincent, Olga A. Bogatyreva and Nikolaj R. Bogatyrev, "Biomimetics: Its Practice and Theory," *Journal of the Royal Society: Interface* 3 (2006): 471–82. doi:10.1098/rsif.2006.0127.

3 **We might add:** Owen Holland and David McFarland, "History of Models in Ethology," in *Artificial Ethology*, eds. Owen Holland and David McFarland (Oxford: Oxford University Press, 2001), 1–14.

3 **Writing in the:** Yoseph Bar-Cohen, "Biomimetics—Using Nature to Inspire Human Innovation," *Bioinspiration & Biomimicry* 1 (2006): 1; Yoseph Bar-Cohen, ed., *Biomimetics: Nature-Based Innovation* (Boca Raton, FL: CRC Press, 2011).

3 **These suggest a mechanistic**: Janine Benyus, *Biomimicry: Innovation Inspired by Nature* (New York: William Morrow, 1997).

3 **"Although it is":** Julian F. V. Vincent and Olga A. Bogatyreva, "Biomimetics: Its Practice and Theory," 471–82.

4 **The vexing question:** You can get a sense of the debate over the mimeticity of being human in Eric Auerbach, *Mimesis: The Representation of Reality in Western Thought*, 1946, Reprint (Princeton, NJ: Princeton University Press, 2016); Richard Rorty, *Philosophy and the Mirror of Nature*, 1979, Reprint (Princeton, NJ: Princeton University Press, 2017); Walter Benjamin, "On the Mimetic Faculty," *Reflections* (New York: Schocken Books, 1986), 333–37; Jacques Derrida, "The Double Question," *Dissemination*, trans. Barbara Johnson (Chicago, IL: University of Chicago Press, 1981), 173–286.

4 **"I call it":** Michael Taussig, *Mimesis and Alterity: A Particular History of the Senses* (London: Routledge, 1993), x; Taussig is invoking Benjamin's use of the term "mimetic faculty" here. See Benjamin, *Reflections*, 333–37.

4 **And yet, as the classicist:** Stephen Halliwell, "Aristotelian Mimesis Reevaluated," *Journal of the History of Philosophy* 28, no. 4 (1990): 487.

4 **In 2007, the literary critic:** Timothy Morton, *Ecology without Nature: Rethinking Environmental Aesthetics* (Cambridge, MA: Harvard University Press, 2007).

4 **They've even:** Antonio Damasio, *Self Comes to Mind: Constructing the Conscious Brain* (New York: Random House, 2010), 110–12.

5 **Seeing an expression:** Vittorio Gallese et al., "Mirror Neuron Forum," *Perspectives on Psychological Science* 6, no. 4 (2011): 369–407; Marco Iacoboni, "Imitation, Empathy, and Mirror Neurons," *Annual Review of Psychology* 60 (2009): 653–70, doi:10.1146/annurev.psych.60.110707.163604; Marco Iacoboni, "Failure to Deactivate in Autism: The Co-Constitution of Self and Other," *TRENDS in Cognitive Sciences* 10, no. 10 (2006): 432.

5 **More broadly, we might:** Stacy Alaimo and Susan Hekman, eds., *Material Feminisms* (Bloomington: Indiana University Press, 2008); Diana Coole and Samantha Frost, eds., *New Materialisms: Ontology, Agency and Politics* (Durham, NC: Duke University Press, 2010); Rodney Brooks, "Elephants Don't Play Chess," *Robotics and Autonomous Systems* 6 (1990): 3–15.

5 **"Mimesis is best understood":** William Schweiker, "Beyond Imitation: Mimetic Praxis in Gadamer, Ricoeur, and Derrida," *The Journal of Religion* 68, no. 1 (1988): 24.

5 **It bears some kinship:** Ian Bogost, *Alien Phenomenology: Or What It's Like to Be a Thing* (Minneapolis: University of Minnesota Press, 2013), 71; Karen Barad, *Meeting the Universe Halfway* (Durham, NC: Duke University Press, 2007).

5 **"Some metaphors *do* fail":** Graham Harman, *Guerilla Metaphysics: Phenomenology and the Carpentry of Things* (Chicago, IL: Open Court Publishing, 2005), 117–24; Barad, *Meeting.*

6 **I'd encountered something:** Eric Higgs, *Nature by Design: People, Natural Process, and Ecological Restoration* (Cambridge: MIT Press, 2003).

7 **Why not learn:** Anna Lowenhaupt Tsing, *The Mushroom at the End of the World: On the Possibility of Life in Capitalist Ruins* (Princeton, NJ: Princeton University Press, 2015).

7 **This resistance is longstanding:** Steven Vogel, *Cats' Paws and Catapults: Mechanical Worlds of Nature and People* (New York: Norton, 1998), 10.

8 **When Bill McKibben published:** Bill McKibben, *The End of Nature* (New York: Random House, 1989).

8 **This struggle to articulate:** Bruno Latour, *Facing Gaia: Eight Lectures on the New Climate Regime* (Medford, MA: Polity Press, 2017); Donna Haraway, *Staying with the Trouble: Making Kin in the Chthulucene* (Durham, NC: Duke University Press, 2016); *Simians, Cyborgs, and Women: The Reinvention of Nature* (London: Routledge, 1990).

9 **Take Joyce Carol Oates':** Joyce Carol Oates, "Against Nature," *Antaeus* 57 (1986): 238–43; Michael Specter, interview with Terry Gross, *Fresh Air*, June 26, 2013.

1. Trailhead: quadrupeds and Plantoids

12 **Or maybe our:** Frank E. Lutz, *Nature Trail: An Experiment in Out-Door Education* (New York: American Museum of Natural History, 1926), Miscellaneous Publications no. 21.

13 **The scope of the journey varies:** Aldo Leopold, *A Sand County Almanac*, (Oxford: Oxford University Press, 1949; Oxford: Oxford University Press, 1992).

13 **Then there's the conservationist:** J. Michael Fay, *Megatransect: Mike Fay's Journals* (Washington, DC: National Geographic, 2005).

13 **There's John Muir:** John Muir, "A Wind-storm in the Forests, 1894," in *Nature Writings: The Story of My Boyhood and Youth; My First Summer in the Sierra; The Mountains of California; Stickeen; Essays* (New York: Library of America, 1997), 234; Henry David Thoreau, "Walking, 1862," in *Thoreau: Collected Essay and Poems* (New York: Library of America, 2001).

13 **The MIT researcher:** Rodney Brooks, "Elephants Don't Play Chess," *Robotics and Autonomous Systems* 6 (1990): 3–15; Hans Moravec, *Mind Children* (Cambridge, MA: Harvard University Press, 1988), 15.

14 **More recently:** Karen Barad, "*Posthumanist Performativity:* Toward an Understanding of How Matter Comes to Matter," *Signs* 28, no. 3 (2003): 829.

16 **One possible result:** Levi R. Bryant, *The Democracy of Objects* (Ann Arbor, MI: Open Humanities Press, 2011) in which he describes this rhizomatic or web-like system of the living and nonliving as objects; also Ian Bogost, *Alien Phenomenology, or What It's Like to Be a Thing* (Minneapolis: University of Minnesota Press, 2012).

16 **Instead all entities:** Gary Snyder, *The Practice of the Wild* (San Francisco, CA: North Point Press, 1990).

16 **We're all objects**: Graham Harman, *Tool-Being: Heidegger and the Metaphysics of Objects* (Chicago, IL: Open Court, 2002) and *Object-Oriented Ontology: A New Theory of Everything* (London: Penguin, 2018).

17 **Biomimetics is a peculiar rejoinder:** Bogost, *Alien*, 50.

17 **Think of avid birders:** Snyder, *Practice*, 14.

21 **As one of the first:** Claudio Semini et al., "Design of HyQ – A Hydraulically and Electrically Actuated Quadruped Robot," *Proceedings of the Institution of Mechanical Engineers, Part I: Journal of Systems and Control Engineering* 225, no. 6 (2011): 831–49.

21 **The evolution in the physical shape:** Rolf Müller et al., "Biodiversifying Bioinspiration," *Bioinspiration & Biomimetics* 13, no. 5 (2018): 053001. The authors argue that bioinspired innovation follows a pattern in which breakthroughs occur as new biological research appears, with subsequently minor advancement as the engineering world waits for new biotic insight.

21 **In 1878, the pioneering:** Eadweard Muybridge. *The Horse in motion. "Sallie Gardner," owned by Leland Stanford; running at a 1:40 gait over the Palo Alto track, 19th June/ Muybridge.* California Palo Alto, ca. 1878. Photograph. https://www.loc.gov/item/97502309/.

22 **Semini mentioned an:** Donald F. Hoyt and C. Richard Taylor, "Gait and the Energetic of Locomotion in Horses," *Nature* 292, no. 5820 (1981): 239–40; Robert J. Full and Claire T. Farley, "Musculoskeletal Dynamics in Rhythmic Systems: A Comparative Approach to Legged Locomotion" in *Biomechanics and Neural Control of Posture and Movement*, eds. J. M. Winters et al. (New York: Springer-Verlag, 2000), 192–205; Michael H. Dickinson et al., "How Animals Move: An Integrative View," *Science* 288, no. 5463 (2000): 100–06.

22 **Instead of observing:** Alan M. Wilson, et al., "Biomechanics of Predator–Prey Arms Race in Lion, Zebra, Cheetah and Impala," *Nature* 554, no. 7691 (2018): 183–88.

23 **One canine version:** Ioannis Poulakakis, "Bioinspired Robotic Quadrupeds," in *Bioinspired Legged Locomotion: Models, Concepts and Applications*, eds. Maziar A. Sharbafi and André Seyfarth (London: Butterworth-Heinemann, 2017), 527–61; Koh Hosoda et al., "Actuation in Legged Locomotion," *Bioinspired Legged Locomotion: Models, Concepts and Applications*, eds. Maziar A. Sharbafi and André Seyfarth (London: Butterworth-Heinemann, 2017), 563–90.

24 **But because it couldn't:** Stéphane Bazeille et al., "Quadruped Robot Trotting over Irregular Terrain Assisted by Stereo-vision, *Intelligent Service Robotics* 7, no. 2 (2014): 67–77; Auke Jan Ijspeert, "Central Pattern Generators for Locomotion Control in Animals and Robots: A Review," *Neural Networks* 21 (2008): 642–53.

25 **With conditions like:** Claudio Semini et al., "Towards Versatile Legged Robots Through Active Impedance Control," *The International Journal of Robotics Research*, 34, no. 7 (2015): 1003–20; Victor Barasuol et al., "A Reactive Controller Framework for Quadrupedal Locomotion on Challenging Terrain," *Proceedings – IEEE International*

Conference on Robotics and Automation (2013), doi:10.1109/ICRA.2013.6630926; Thiago Boaventura, Gustavo A. Medrano-Cerda and Claudio Semini, "Stability and Performance of the Compliance Controller of the Quadruped Robot HyQ," *IEEE/RSJ International Conference on Intelligent Robots and Systems* (2013), doi:10.1109/iros.2013.6696541.

29 **Trees, he argued:** Hans Moravec, "Locomotion, Vision and Intelligence," in *Robotics Research – The First International Symposium,* eds. Michael Brady and Richard Paul (Cambridge: MIT Press, 1984), 215–24.

29 **Plants, they contend:** Barbara Mazzolai et al., "The Plant as a Biomechatronic System," *Plant Signaling & Behavior* 5, no. 2 (2010): 90–3.

30 **It turns out:** Robert J. Ferl and Anna-Lisa Paul, "The Effect of Spaceflight on the Gravity-sensing Auxin Gradient of Roots: GFP Reporter Gene Microscopy on Orbit," *Microgravity* (2016), doi:10.1038/npjmgrav.2015.23.

30 **When a root hits:** Shahzad Zaigham et al., "A Potassium-dependent Oxygen Sensing Pathway Regulates Plant Root Hydraulics," *Cell* 167, no. 1 (2016): 87–98.

30 **The root cap itself:** Emanuela Del Dottore et al., "An Efficient Soil Penetration Strategy for Explorative Robots Inspired by Plant Root Circumnutation Movements," *Bioinspiration & Biomimicry* 13 (2018): 015003; Ali Sadeghi et al, "A Novel Growing Device Inspired by Plant Root Soil Penetration Behaviors," *PLoS ONE* 9, no. 2 (2014): e90139.

32 **Del Dottore uncapped:** Ali Sadeghi et al., "A Plant-Inspired Robot with Soft Differential Bending Capabilities," *Bioinspiration & Biomimetics* 12 (2016): 015001, doi:10.1088/1748-3190/12/1/015001.

32 **"If you are rigid":** Barbara Mazzolai, "Plant-inspired growing robots," *Soft Robotics: Trends, Applications and Challenges* (Berlin: Springer, 2017), 57–63; Cecilia Laschi, Barbara Mazzolai and Matteo Cianchetti, "Soft Robotics: Technologies and Systems Pushing the Boundaries of Robot Abilities," *Science Robotics* 1 (2016), doi:10.1126/scirobotics.aah3690.

32 **She showed me:** Ali Sadeghi, A Mondini and Barbara Mazzolai, "Towards Self-Growing Soft Robots Inspired by Plant Roots and Based on Additive Manufacturing Technologies," *Soft Robotics* 4, no. 3 (2017): 211–23, doi:10.1089/soro.2016.0080.

33 **Mechanisms like these:** Silvia Taccola et al., "Toward a New Generation of Electrically Controllable Hygromorphic Soft Actuators," *Advanced Materials* 27, no. 10 (2015): 1668–75.

34 **As Mazzolai put it:** Frank Veenstra et al., "Toward Energy Autonomy in Heterogeneous Modular Plant-Inspired Robots through Artificial Evolution," *Frontiers in Robotics and AI* 4 (2017), doi:10.3389/frobt.2017.00043.

34 **Del Dottore had mentioned:** I also spoke with Elliot Hawkes at the University of California, Santa Barbara, whose biomechanics lab is developing soft pneumatic actuators based on the mechanics of growing vines. See Elliot W. Hawkes et al., "A Soft Robot that Navigates Its Environment through Growth," *Science Robotics* 2, no. 8 (2017): eaan3028.

2. Interpretive station: bats and biomimetic sensors

38 **It seems everybody**: Thomas Nagel, "What Is it Like to Be a Bat?" *The Philosophical Review* 83, no. 4 (1974): 435–50; J.M. Coetzee, *Elizabeth Costello* (New York: Viking, 2003); Merlin Sheldrake, *Entangled Life: How Fungi Make Our World, Change Our Minds and Make Our Futures* (New York: Random House, 2020), 3; Charles Foster, *Being a Beast* (New York: Metropolitan Books, 2016); Jenni Diski, *What I Don't Know About Animals* (New Haven, CT: Yale University Press, 2011); Thomas Thwaites, *Goatman: How I Took a Holiday from Being Human* (Princeton, NJ: Princeton Architectural Press, 2016); See also, Eduardo Navarro, "Timeless Alex," 2015, https://www.newmuseum.org/calendar/view/464/eduardo-navarro-1.

38 **"Even Without"**: Nagel, "What Is it Like to Be a Bat?" 438.
38 **"It will not help"**: Ibid., 439.
38 **As Bogost puts it**: Ian Bogost, *Alien Phenomenology, or What It's Like to Be a Thing* (Minneapolis: University of Minnesota Press, 2012).
39 **As biomimetic practitioners**: Coetzee, *Elizabeth Costello*, 80.
40 **We like to think**: Timothy Morton, *Hyperobjects: Philosophy and Ecology after the End of the World* (Minneapolis: University of Minnesota Press, 2013).
41 **In reality, of course**: Michael J. Harvey, J. Scott Altenbach and Troy L. Best, *Bats of the United States and Canada* (Baltimore, MA: Johns Hopkins University Press, 2006); Ronald M. Nowak, *Walker's Bats of the World* (Baltimore, MA: Johns Hopkins University Press, 1994).
41 **We now have galleries**: "Species Profiles," Bat Conservation International, http://www.batcon.org/resources/media-education/species-profiles.
42 **One answer lies**: John D. Altringham, *Bats: From Evolution to Conservation* (Oxford: Oxford University Press, 2011), 61–86.
42 **Echolocation is a call**: M. Brock Fenton, "Evolution of Echolocation," in *Bat Evolution, Ecology, and Conservation*, eds. Rick A. Adams and Scott C. Pedersen (New York: Springer, 2013),: 47–70.
43 **Many horseshoe bats**: Lasse Jakobsen, Signe Brinkløv and Annemarie Surlykke, "Intensity and Directionality of Bat Echolocation Signals," *Frontiers in Physiology* 4 (2013), https://doi.org/doi:10.3389/fphys.2013.00089.
43 **Comparative studies of related moths**: James H. Fullard, "Acoustic Relationships Bbetween Tympanate Moths and the Hawaiian Hoary Bat (Lasiurus cinereus semotus), *Journal of Comparative Physiology A* 155 (1984): 795–801, https://doi.org/doi:10.1007/BF00611596.
44 **A moth may turn**: Gareth Jones and Jens Rydell, "Attack and Defense: Interactions Bbetween Bats and Their Insect Prey," in *Bat Ecology,* eds. Thomas H. Kunz and M. Brock Fenton (Chicago, IL: University of Chicago Press, 2003), 301–332; Lee A. Miller and Annemarie Surlykke, "How Some Insects Detect and Avoid Being Eaten by Bats: Tactics and Countertactics of Prey and Predator: Evolutionarily Speaking, Insects Have Responded to Selective Pressure from Bats with New Evasive Mechanisms, and These Very Responses in Turn Put Pressure on Bats to "Improve" Their Tactics," *BioScience* 51, no. 7 (2001): 570–581.
44 **While it might seem**: Gareth Jones and Emma C. Teeling, "The Evolution of Echolocation in Bats," *TRENDS in Ecology and Evolution,* 21, no. 3 (2006): 149–156.
44 **The differences in bat facial features**: Sharon M. Swartz, Patricia W. Freeman and Elizabeth F. Stockwell, "Ecomorphology of Bats: Comparative and Experimental Approaches Relating Structural Design to Ecology," in *Bat Ecology,* eds. Thomas H. Kunz and M. Brock Fenton (Chicago, IL: University of Chicago Press, 2003), 257–292.
44 **Some species broadcast**: Qiao Zhuang and Rolf Müller, "Noseleaf Furrows in a Horseshoe Bat Act as Resonance Cavities Shaping the Biosonar Beam," *Physical Review Letters* 97, no. 21 (2006): 218701-–04.
44 **Müller's research focuses**: Rolf Müller et al., "Biodiversifying Bioinspiration," *Bioinspiration & Biomimetics* 13, no. 5 (2018): 053001.
45 **If a conventional**: Bryan D. Todd and Rolf Müller, "A Comparison of the Role of Beamwidth in Biological and Engineered Sonar," *Bioinspiration & Biomimicry* 13, no. 1 (2018), https://iopscience.iop.org/article/10.1088/1748-3190/aa9a0f/meta; Philip Caspers and Rolf Müller, "A Design for a Dynamic Biomimetic Sonarhead Inspired by Horseshoe Bats," *Bioinspiration & Biomimicry* 13, no. 4 (2018): 046011; Luhui Yang, Allison Yu and Rolf Müller, "Design of a Dynamic Sonar Emitter Inspired by Hipposiderid Bats," *Bioinspiration & Biomimetics* 13, no. 5 (2018): 056003; Beatrice D. Lawrence and James A. Simmons, "Echolocation in Bats: The External Ear and Perception of the Vertical Positions of Targets," *Science,* 218, no. 4571 (1982): 481–83; James A. Simmons et al., "Acuity of Horizontal Angle Discrimination by the Echolocating Bat, *Eptesicus fuscus*," *Journal of Comparative Physiology* 153, no. 3 (1983): 321–30.

45 **This work has taken**: Jianguo Ma and Rolf Müller, "A Method for Characterizing the Biodiversity in Bat Pinnae as a Basis for Engineering Analysis," *Bioinspiration & Biomimicry* 6, no. 2 (2011): 026008; See also Müller, "Biodiversifying Bioinspiration," which makes the same point about adaptive radiation as technological principle.

47 **In fact, we know**: Chen Ming, Hongxiao Zhu and Rolf Müller, "A Simplified Model of Biosonar Echoes from Foliage and the Properties of Natural Foliages," *PLoS ONE* 12, no. 12 (2017doi:10.1371/journal. pone.0189824.

47 **By providing a safe**: Michael G. Schöner et al., "Bats Are Acoustically Attracted to Mutualistic Carnivorous Plants," *Current Biology* 25, no. 14 (2015): 1911–16.

48 **Recent models then**: Aristid Lindenmayer, "Mathematical Models for Cellular Interaction in Development, Parts I and II," *Journal of Theoretical Biology* 18 (1968): 280–315.

48 **The first of these alfresco**: Chen Ming, *Foliage Echoes and Sensing in Natural Environments*, PhD diss., Virginia Tech, 2017, https://vtechworks.lib.vt.edu/bitstream/handle/10919/78825/Ming_C_D_2017.pdf.

54 **To address these questions**: Susan C. Loeb et al., *A Plan for the North American Bat Monitoring Program (NABAT),* General Technical Report SRS-208 (Asheville, NC: U.S. Department of Agriculture Forest Service, Southern Research Station, 2015).

55 **It was first recognized**: Craig L. Frank, April D. Davis, and Carl Herzog, "The Evolution of a Bat Population with White-Nose Syndrome (WNS) Reveals a Shift from an Epizootic to an Enzootic Phase," *Frontiers in Zoology* 16, no. 40 (2019), doi:10.1186/s12983-019-0340-y; Gregory G. Turner, DeeAnn M. Reeder, and Jeremy T. H. Coleman, "A Five-year Assessment of Mortality and Geographic Spread of White-Nose Syndrome in North American Bats, with a Look at the Future," *Bat Research News,* 52 (2011): 13–27.

56 **Mostly, these were**: Thomas H. Kunz and Linda Lumsden, "Ecology of Cavity and Foliage Roosting Bats," in *Bat Ecology,* eds. Thomas H. Kunz and M. Brock Fenton (Chicago, IL: University of Chicago Press, 2003), 3–69; U.S. Fish & Wildlife Service, "Tricolored Bat," *Conserving South Carolina's At-Risk Species: Species Facing Threats to Their Survival,* February 2019, https://www.fws.gov/southeast/pdf/fact-sheet/tri-colored-bat.pdf.

59 **They'd burned through**: Michelle L. Verant et al., "White-nose Syndrome Initiates a Cascade of Physiologic Disturbances in the Hibernating Bat Host," *BMC Physiology* 14, no. 1 (2014): 10; DeeAnn M. Reeder et al., "Frequent Arousal from Hibernation Linked to Severity of Infection and Mortality in Bats with White-Nose Syndrome," *PLoS ONE* 7, no. 6 (2012): e38920.

62 **He mentioned one**: Pauline Oliveros, *Deep Listening: A Composer's Sound Practice* (Bloomington, IN: iUniverse, 2005).

62 **De Little also described**: Salomé Voegelin, *The Political Possibility of Sound: Fragments of Listening* (London: Bloomsbury Academic, 2018) and *Listening to Noise and Silence: Towards a Philosophy of Sound Art* (London: Continuum, 2010); Brandon Labelle, *Sonic Agency: Sound and Emergent Forms of Resistance* (London: Goldsmiths Press/Sonics Series, 2018); Stephen Feld, "Acoustemology," in *Keywords in Sound,* eds. David Novak and Matt Sakakeeny (Durham, NC: Duke University Press, 2016), 12–21.

64 **We continued with this sound**: Alex De Little, *Spatial Listening: Exercises in Aural Architecture* (London: Wild Pansy Press, 2019).

64 **De Little has described**: De Little, *Spatial Listening*, 3.

3. Interpretive station: social insects and robotic swarms

68 **Choosing to build**: Stephen C. Pratt, "Collective Control of the Timing and Type of Comb Construction by Honey Bees (*Apis mellifera*)," *Apidologie* 35 (2004): 193–205.

69 **Complex behavior like comb**: Sara Brin Rosenthal et al., "Revealing the Hidden Networks of Interaction in Mobile Animal Groups Allows Prediction of Complex Behavioral Contagion," *PNAS* 112, no. 15 (2015): 4690–95.

69 **What is wildness**: Gary Snyder, *The Practice of the Wild* (San Francisco, CA: North Point Press, 1990), 9–10.
69 **An "ordering of impermanence"**: Snyder, *Practice,* 5.
70 **The famed ant biologists**: Edward O. Wilson and Bert Hölldobler, *The Ants* (Cambridge, MA: Harvard University Press, 1990).
70 **It's self-abnegation**: Bert Hölldobler and Edward O. Wilson, *The Superorganism: The Beauty, Elegance, and Strangeness of Insect Societies* (New York: Norton, 2009).
71 **Dorigo proposed**: Marco Dorigo, Mauro Birattari, and Thomas Stützle, "Ant Colony Optimization: Artificial Ants as a Computational Intelligence Technique," *IEEE Computational Intelligence Magazine* 1, no. 4 (2006): 28–39; Marco Dorigo and Thomas Stützle, *Ant Colony Optimization* (Cambridge: MIT Press, 2004).
72 **In essence, the colony**: Jean-Louis Deneubourg et al., "The Self-Organizing Exploratory Pattern of the Argentine Ant," *Journal of Insect Behavior* 3, no. 2 (1990): 159–68. The article opens by citing the ant colony as superorganism concept developed by Hölldobler and Wilson as inspiration for describing emergent collective behavior.
72 **Instead, Dorigo's model**: Dorigo, "Ant Colony Optimization," 30–31.
72 **The physicist Paul C.W. Davies**: Paul C.W. Davies, "Preface," in *The Re-Emergence of Emergence: The Emergentist Hypothesis from Science to Religion,* eds. Paul Davies and Philip Clayton (Oxford: Oxford University Press, 2006), x.
73 **"If there are phenomena"**: David J. Chalmers, "Strong and Weak Emergence," in *The Re-Emergence of Emergence: The Emergentist Hypothesis from Science to Religion,* eds. Paul Davies and Philip Clayton (Oxford: Oxford University Press, 2006), 244–56.
73 **However, what feels**: Mark A. Bedau, "Weak Emergence," in *Philosophical Perspectives: Mind, Causation, and World*, ed. James Tomberlin (Malden, MA: Blackwell, 1997), 375–99.
73 **Yet the discovery**: Jane Bennett, "A Vitalist Stopover on the Way to a New Materialism," in *New Materialisms: Ontology, Agency and Politics*, eds. Diana Coole and Samantha Frost (Durham, NC: Duke University Press, 2010), 47–69.
73 **One minute a host**: Jerome Buhl et al., "From Disorder to Order in Marching Locusts," *Science* 312 (2006): 1402–06; Sepideh Bazazi et al., "Collective Motion and Cannibalism in Locust Migratory Bands," *Current Biology* 18, no. 10 (2008): 735–39.
74 **It runs through**: Jack Kerouac, *The Dharma Bums* (London: Penguin, 1976), 157.
74 **We can encounter it**: Edward Abbey, *The Journey Home* (New York: E.P. Dutton, 1977), 22.
74 **The difference between**: Morton offers a corollary distinction between "strong" and "weak" ecomimesis, although he suggests that both are ultimately untenable. The strong is "magical," and no mere copy. Morton, *Ecology*, 54.
75 **Instead of the hallmarks**: Diana Coole and Samantha Frost, "Introducing the New Materialisms," in *New Materialisms: Ontology, Agency and Politics*, eds. Diana Coole and Samantha Frost (Durham, NC: Duke University Press, 2010), 14.
75 **Barad's point takes**: Karen Barad, "Posthumanist Performativity: Toward an Understanding of How Matter Comes to Matter," in *Material Feminisms*, eds. Stacy Alaimo and Susan Hekman (Bloomington: Indiana University Press, 2008), 139.
75 **She wants things**: Karen Barad, *Meeting the Universe Halfway: Quantum Physics and the Entanglement of Matter and Meaning* (Durham, NC: Duke University Press, 2007), 132.
76 **This emphasis on**: Snyder, *Practice*, 10.
76 **Barad's skepticism**: Barad, *Meeting*, 132.
76 **And soon after**: Barad, *Meeting*, 382.
79 **Pratt has put**: Stephen Pratt, "Tandem Recruitment by Emigrating Ants," https://www.youtube.com/watch?v=Kam0jKiAk3c.
79 **When *Temnothorax albipennis***: Stephen C. Pratt et al., "Quorum Sensing, Recruitment, and Collective Decision-Making During Colony Emigration by the Ant *Leptothorax albipennis*," *Behavioral Ecology and Sociobiology* 52 (2002): 117–27, doi:10.1007/s00265-002-0487-x.

80 **These were desert**: Aurélie Buffin and Stephen C. Pratt, "Cooperative Transport by the Ant *Novomessor cockerelli*," *Insectes Sociaux* 63 (2016), doi:10.1007/s00040-016-0486-y.

81 **Imagine a scenario**: Spring Berman et al., "Experimental Study and Modeling of Group Retrieval in Ants as an Approach to Collective Transport in Swarm Robotic Systems," *Proceedings of the IEEE* 99, no. 9 (2011): 1471.

81 ***N. cockerelli* are known**: Berman et al., "Experimental Study and Modeling," 1472.

85 **He might, for example**: Jean-Pierre de la Croix and Magnus Egerstedt, "Group-Size Selection for a Parameterized Class of Predator-Prey Models," https://magnus.ece.gatech.edu/Papers/LionsMTNS14.pdf.

86 **Just as constructing**: Frank Grasso, "Case Study 1: How Robotic Lobsters Locate Odour Sources in Turbulent Water," in *Artificial Ethology,* eds. Owen Holland and David McFarland (Oxford: Oxford University Press, 2001), 47–59.

87 **On a video posted**: Clint Penick, "Ants Dueling," YouTube, last revised May 9, 2014, https://www.youtube.com/watch?v=74ruqyOUX-8. Penick's research has uncovered the hormonal changes that result from successfully transitioning from worker to gamergate, as well as the chemical surge in dopamine that comes with winning these duels.

87 **A set of probabilities**: Takao Sasaki et al., "A Simple Behavioral Model Predicts the Emergence of Complex Animal Hierarchies," *The American Naturalist* 187, no. 6 (2016): 765–75.

89 **The two groups merged**: Vincent Fourcassié, Audrey Dussutour and Jean-Louis Deneubourg, "Ant Traffic Rules," *Journal of Experimental Biology* 213 (2010): 2357–63.

4. Interpretive station: social robots and other trailside companions

93 **Putting humans into**: Masahiro Mori, "The Uncanny Valley [From the Field]," trans. K. F. MacDorman and N. Kageki, *IEEE Robotics & Automation Magazine* 19, no. 2 (2012): 98–100, doi:10.1109/MRA.2012.2192811.

94 **While there is**: Daniel T. Simon, et al. "An Organic Electronic Biomimetic Neuron Enables Auto-regulated Neuromodulation," *Biosensors and Bioelectronics* 71 (2015): 359–64.

94 **An artificial neural**: John D. Kelleher, *Deep Learning* (Cambridge: MIT Press, 2019): 126–33.

95 **These are incremental**: Kelleher, *Deep Learning*, 65–67; David Watson, "The Rhetoric and Reality of Anthropomorphism in Artificial Intelligence," *Minds and Machines* 29 (2019): 417–40.

95 **Some experts in AI**: Gary Marcus, *"Deep Learning: A Critical Appraisal," ArXiv abs/1801.00631* (2018): n.p.; Rodney Brooks, "Steps Toward Super Intelligence I, How We Got Here." rodneybrooks.com/forai-steps-toward-super-intelligence-i-how-we-got-here/.

95 **There's Annie Dillard**: Annie Dillard, "Living Like Weasels," in *Teaching a Stone to Talk* (New York: Harper Collins, 1998), 65–71; Aldo Leopold, *Sand.*

96 **We could join**: John Muir, "Among the Birds of the Yosemite," in John Muir, *Our National Parks*, 1901, Reprint (Madison, WI: University of Wisconsin Press, 1982), 213–240.

96 **Or Edward**: Edward Abbey, "The Serpents of Paradise," in *Desert Solitaire: A Season in the Wilderness* (New York: Ballentine, 1968; New York: Ballentine, 1985), 17–25.

96 **In fact, eye contact**: Antonia F. de C. Hamilton, "Gazing at Me: The Importance of Social Meaning in Understanding Direct-Gaze Cues, *Philosophical Transactions of the Royal Society of London. Series B, Biological Sciences* 371, no. 1686 (2016), doi:10.1098/rstb.2015.0080; Nathalie George and Laurent Conty, "Facing the Gaze of Others," *Clinical Neurophysiology* 38, no. 3 (2008): 197–207, doi:10.1016/j.neucli.2008.03.001.

96 **Recently, facial recognition**: Oskar Palinko et al., "Eye Gaze Tracking for a Humanoid Robot," *IEEE-RAS 15th International Conference on Humanoid Robots* (2015):

318–24, doi:10.1109/HUMANOIDS.2015.7363561; Takayuki Todo, "SEER: Simulative Emotional Expression Robot," http://www.takayukitodo.com/.

97 **Derrida tells**: Jacques Derrida, "The Animal That Therefore I Am (More to Follow)," *Critical Inquiry* 28, no. 2 (2002): 369–418.

97 **Could we divide**: Herbert Feigl, "Some Crucial Issues of Mind-Body Monism," *Synthese* 22, no. 3/4 (1971): 295–312; Paul D. MacLean, *The Triune Brain in Evolution: Role in Paleocerebral Functions* (New York: Springer, 1990).

98 **Haraway is perhaps**: Donna Haraway, *Simians, Cyborgs, and Women: The Reinvention of Nature* (London: Routledge, 1990).

98 **More recently, however**: Donna Haraway, *When Species Meet* (Minneapolis: University of Minnesota Press, 2008).

98 **In describing the process**: Donna Haraway, *Staying with the Trouble: Making Kin in the Chthulucene* (Durham, NC: Duke University Press, 2016), 58–98.

98 **The concept of autopoiesis**: Francisco Varela, Humberto Maturana, and Rodrigo Uribe, "Autopoiesis: The Organization of Living Systems, Its Characterization, and a Model," *Biosystems* 5 (1974): 559–69; Humberto Maturana and Francisco Varela, *Autopoiesis and Cognition: The Realization of the Living* (Boston, MA: Reidel, 1980).

99 **Empathy is one result**: Iacoboni, "Imitation, Empathy, and Mirror Neurons," 653–70; Iacoboni, "Failure to Deactivate in Autism: The Co-Constitution of Self and Other," 431–33. Antonia F. de C. Hamilton, "Reflecting on the Mirror Neuron System in Autism: A Systematic Review of Current Theories," *Developmental Cognitive Neuroscience* 3 (2013): 92.

99 **The work of**: Antonio Damasio, *Looking for Spinoza: Joy, Sorrow and the Feeling Brain* (New York: Harcourt, 2003), 6.

100 **It does so via**: Damasio, *Self*, 110–12.

99 **Damasio's work offers**: Lorenzo Cominelli, Daniele Mazzei and Danilo Emilio De Rossi, "SEAI: Social Emotional Artificial Intelligence Based on Damasio's Theory of Mind," *Frontiers in Robotics and AI* 5, no. 6 (2018), doi:10.3389/frobt.2018.00006.

99 **These elements of**: Antonio Damasio, *The Strange Order of Things: Life, Feeling and the Making of Cultures* (New York: Pantheon, 2018).

100 **"Rather than up-armouring"**: Kingston Man and Antonio Damasio, "Homeostasis and Soft Robotics in the Design of Feeling Machines," *Nature Machine Intelligence* 1 (2019): 446–52, doi:10.1038/s42256-019-0103-7.

100 **In a recent article**: Erin Hecht et al., "Significant Neuroanatomical Variation among Domestic Dog Breeds," *The Journal of Neuroscience,* 39, no. 39 (2019): 7748, doi:10.1523/JNEUROSCI.0303-19.2019.

100 **Hecht's lab used MRI**: Hecht, "Significant Neuroanatomical Variation among Domestic Dog Breeds," 7756.

101 **This sympoietic self-making**: Haraway makes this connection to Barad's theory explicitly. Donna Haraway, *When Species Meet: Narrating Across Species Lines* (Minneapolis: University of Minnesota Press, 2008), 17. Damasio, however, is dismissive of quantum physics as a model for the mind, a theory whose logic he paraphrases as a coincidence of mysterious phenomena: "The conscious mind seems mysterious; because quantum physics remains mysterious, perhaps the two mysteries are connected." See Antonio Damasio, *Self*, 15.

101 **"Nothing makes"**: Haraway, *Staying,* 58.

101 **Sympoiesis "enfolds"**: Ibid., 58.

101 **The connection**: Vittorio Gallese et al., "Mirror Neuron Forum," *Perspectives on Psychological Science* 6, no. 4 (2011): 369–407; Caroline Catmur, Vincent Walsh and Cecilia Heyes, "Sensorimotor Learning Configures the Human Mirror System," *Current Biology* 17 (2007): 1527–31, doi:10.1016/j.cub.2007.08.006.

102 **Two glowing "eyes"**: Masahira Fujita and Hiroaki Kitano, "Development of an Autonomous Quadruped Robot for Robot Entertainment," *Autonomous Robots* 5 (1998): 7–20, doi:10.1023/A:1008856824126.

102 **If we stopped here**: Fujita and Kitano, "Development of an Autonomous Quadruped Robot for Robot Entertainment," 7.

103 **An "angry" MUTANT**: Ibid., 14–5.

104 **Descartes recognized**: René Descartes, *Discourse on Method and Related Writings,* trans. Desmond C. Clarke (1637; London: Penguin, 1999), 40–1.

104 **I'm reminded of**: David Quammen, "The Face of a Spider," in *The Flight of the Iguana* (New York: Touchstone, 1988), 3–9.

106 **The operating manual**: *AIBO Operation Manual* (Tokyo: The Sony Corporation, 1999), 54.

106 **But AIBO was**: *Raising AIBO—The Handbook* (Tokyo: The Sony Corporation, 1999), 8.

107 **"We strongly believe"**: Fujita and Kitano, "Development of an Autonomous Quadruped Robot for Robot Entertainment," 7.

109 **It's this kind**: Sherry Turkle, *Alone Together: Why We Expect More from Technology and Less from Each Other* (New York: Basic Books, 2011), 43.

110 **"Our eyes locked"**: Dillard, *Teaching,* 66.

110 **Quammen writes**: Quammen, *Flight,* 9.

111 **The Americans with Disabilities**: U.S. Department of Justice, Civil Rights Division, "Service Animals," 2010, https://www.ada.gov/service_animals_2010.htm.

112 **Socially assistive robots**: Thorsten Kolling et al., "What is Emotional about Emotional Robotics?" in *Emotions, Technology and Health,* eds. Sharon Tettegah and Yolanda Garcia (Amsterdam: Elsevier, 2016), 85–103; Adam Miklosi and Marta Gacsi, "On the Utilization of Social Animals as a Model for Social Robotics," *Frontiers in Psychology* 3, no. 75 (2012): doi:10.3389/fpsyg.2012.00075.

112 **When Takanori Shibata**: Kazuyoshi Wada and Takanori Shibata, "Living With Seal Robots-Its Sociopsychological and Physiological Influences on the Elderly at a Care House," *IEEE Transactions on Robotics* 23, no. 5 (2007): 972–80, doi:10.1109/TRO.2007.906261.

112 **Researchers have also**: Andrea Beetz et al., "Psychosocial and Psychophysiological Effects of Human-Animal Interactions: The Possible Role of Oxytocin," *Frontiers in Psychology* 3, no. 234 (2012), doi:10.3389/fpsyg.2012.00234.

113 **In the video**: "A Real Seal," PARO, http://www.parorobots.com/video.asp.

113 **In fact, one field**: Kit M. Kovacs, "Maternal Behavior and Early Behavioral Ontogeny of Harp Seals, *Phoca groenlandica,*" *Animal Behavior* 35, no. 3 (1987): 849, doi:10.1016/S0003-3472(87)80120-3.

113 **A creature that**: Ilse C. Van Opzeeland and Sofie M. Van Parijs, "Individuality in Harp Seal, *Phoca groenlandica,* Pup Vocalizations," *Animal Behaviour* 68 (2004): 1115–23, doi:10.1016/j.anbehav.2004.07.005.

115 **In designing the robot**: Selma Šabanović et al., "PARO Robot Affects Diverse Interaction Modalities in Group Sensory Therapy for Older Adults with Dementia," *IEEE/International Conference on Rehabilitation Robotics* (2013): 2, doi:10.1109/ICORR.2013.6650427.

115 **They talked to each other**: Hayley Robinson et al., "The Psychosocial Effects of a Companion Robot: A Randomized Controlled Trial," *JAMDA* 14, no. 9 (2013): 661–7, doi:10.1016/j.jamda.2013.02.007.

115 **Some of these benefits**: Šabanović, "PARO Robot Affects Diverse Interaction Modalities in Group Sensory Therapy for Older Adults with Dementia," 5–6.

117 **That's clearly not**: Jodi Forlizzi, Carl DiSalvo, and Francine Gemperle, "Assistive Robotics and an Ecology of Elders Living Independently in Their Homes," *Human-Computer Interaction* 19, no. 1 (2004): 25–59, doi:10.1080/07370024.2004.9667339; for broader context see Wan-Ling Chang, Selma Šabanović and Lesa Huber, "Situated Analysis of Interactions between Cognitively Impaired Older Adults and the Therapeutic Robot PARO," in *Social Robotics, ICSR 2013,* eds. Guido Herrmann et al. (Cham: Springer, 2013), 371–80, doi:10.1007/978-3-319-02675-6_37.

5. Trail's end: burgers and biomimetic bodies

122 **Before we left**: Sauli Laaksonen, Jyrki Pusenius, Jouko Kumpula, "Climate Change Promotes the Emergence of Serious Disease Outbreaks of Filarioid Nematodes," *EcoHealth* 7, no. 1 (2010): 7–13, doi:10.1007/s10393-010-0308-z.

123 **One, venerable**: Paul Shepard, *Coming Home to the Pleistocene* (Washington, DC: Island Press, 1998), 170.

123 **I tend to think**: Elina Helander-Renvall, "Logical Adaptation to Modern Technology: Snowmobile Revolution in Sápmi," in *The Borderless North*, eds. Lassi Heininen and Kari Laine (Oulu: The Thule Institute, 2008), 27–33.

125 **Whether it's**: Ole Henrik Magga, "Diversity in Saami Terminology for Reindeer, Snow, and Ice," *International Social Science Journal* 58, no. 187 (2006): 25–34, doi:10.1111/j.1468-2451.2006.00594.x; Inger Marie Eira et al., "Traditional Sámi Snow Terminology and Physical Snow Classification—Two Ways of Knowing," *Cold Regions Science and Technology* 85 (2012), doi:10.1016/j.coldregions.2012.09.004.

126 **The anonymous courtier**: Ian Whitaker, "Ohthere's Account Reconsidered," *Arctic Anthropology* 18, no. 1 (1981): 1–11; Julia Fernandez Cuesta, Inmaculata Senra Silva, "Ohthere and Wulfstan: One or Two Voyagers at the Court of King Alfred?" *Studio Neophilologica* 72, no. 1 (2000):18–23; doi:10.1080/003932700750041568.

128 **There were reindeer**: Terhi Vuolaja-Magga et al., "Resonance Strategies of Sami Reindeer Herders in Northernmost Finland during Climatically Extreme Years," *Arctic* 64, no. 2 (2011): 227–41.

130 **For several years**: Alfred Colpaert, Jouko Kumpula and Mauri Nieminen, "Reindeer Pasture Biomass Assessment Using Satellite Remote Sensing," *Arctic* 56 (2003): 147–58, doi:10.14430/arctic610; Jouko Kumpula, Stéphanie Lefrère and Mauri Nieminen, "The Use of Woodland Lichen Pasture by Reindeer in Winter with Easy Snow Conditions," *Arctic* 57 (2004), doi:10.14430/arctic504; Mauri Nieminen, "Why Supplementary Feeding in Finland?" *Rangifer Report* 14 (2010): 4041; Heidi Kitti et al., "Defining the Quality of Reindeer Pastures: The Perspectives of Sámi Reindeer Herders," in *Reindeer Management in Northernmost Europe: Linking Practical and Scientific Knowledge in Social-Ecological Systems*, eds. Bruce C. Forbes et al. (Cham: Springer, 2006), 141–65.

131 **Before there were**: Terhi Vuolaja-Magga et al. "Resonance Strategies of Sami Reindeer Herders in Northernmost Finland during Climatically Extreme Years," *ARCTIC* 64 (2011): 231–32. doi: 10.14430/arctic4102.

133 **"This case description"**: Boris Kan et al., "Suspected Lice Eggs in the Hair of a Boy Revealed Dangerous Parasite," *Lakartidningen* 107, no. 26–28 (2010):1694–97; Jörgen Landehag et al., "Human Myiasis Caused by the Reindeer Warble Fly, *Hypoderma tarandi*, Case Series from Norway, 2011 to 2016," *EuroSurveillance* 22, no. 29 (2017): 30576. doi:10.2807/1560-7917.ES.2017.22.29.30576.

134 **Antlers are**: Tim H. Clutton-Brock, "The Functions of Antlers," *Behaviour* 79, no. 2/4 (1982): 108–25; Gerald Lincoln, "Biology of Antlers," *Journal of Zoology* 226 (1992): 517–28; doi:10.1111/j.1469-7998.1992.tb07495.x.

135 **Each reindeer owner**: Hugh Beach, "Reindeer Ears: Calf Marking during the Contemporary Era of Extensive Herding in Swedish Saamiland," *Annales Societatis Litterarum Humaniorum Regiae Upsaliensis* (2008), 91–118.

136 ***Rangifer* are**: Zeshan Lin et al., "Biological Adaptations in the Arctic Cervid, the Reindeer (*Rangifer tarandus*), *Science* 364, no. 6446 (2019): 1154, doi:10.1126/science.aav6312.

137 **Cultured meat is**: Mark J. Post, "Cultured Meat from Stem Cells: Challenges and Prospects," *Meat Science* 92, no. 3 (2012): 297–301, doi:10.1016/j.meatsci.2012.04.008; Neil Stephens et al., "Bringing Cultured Meat to Market: Technical, Socio-Political, and Regulatory Challenges in Cellular Agriculture," *Trends in Food Science & Technology* 78 (2018): 155–66, doi:10.1016/j.tifs.2018.04.010.

140 **"Our Impossible Burger"**: Neil Stephens, "Growing Meat in Laboratories: The Promise, Ontology and Ethical Boundary-Work of Using Muscle Cells to Make Food," *Configurations: A Journal of Literature, Science and Technology* 21, no. 2 (2013): 159–83, doi:10.1353/con.2013.0013; Erik Jönsson, Tobias Linné, and Ally McCrow-Young, "Many Meats and Many Milks? The Ontological Politics of a Proposed Post-Animal Revolution," *Science as Culture* 28, no. 1 (2019): 70–97, doi:10.1080/09505431.2018.1544232.

140 **This was the one**: Rachel Z. Fraser et al., "Safety Evaluation of Soy Leghemoglobin Protein Preparation Derived from *Pichia pastoris*, Intended for Use as a Flavor Catalyst in Plant-Based Meat," *International Journal of Toxicology*, 37, no. 3 (2018): 241–62, doi:10.1177/1091581818766318.

140 **His research team**: Ross C. Hardison, "A Brief History of Hemoglobins: Plant, Animal, Protist, and Bacteria," *PNAS* 93, no. 12 (1996): 5675–79, doi:10.1073/pnas.93.12.5675; Carol R. Andersson et al., "A New Hemoglobin Gene from Soybean: A Role for Hemoglobin in All Plants," *PNAS* 93, no. 12 (1996): 5682–87, doi:10.1073/pnas.93.12.5682.

143 **Impossible Foods makes**: Pierre J. Gerber et al., *Tackling Climate Change Through Livestock–A Global Assessment of Emissions and Mitigation Opportunities* (Rome: Food and Agriculture Organization of the United Nations, 2013); Henning Steinfeld et al., *Livestock's Long Shadow: Environmental Issues and Options* (Rome: Food and Agriculture Organization of the United Nations, 2006); Mario Herrero et al., "Biomass Use, Production, Feed Efficiencies, and Greenhouse Gas Emissions from Global Livestock Systems," *PNAS* 110, no.52 (2013): 20888–93, doi:10.1073/pnas.1308149110.

143 **The Dutch philosopher**: Cor van der Weele and Clemens Driessen, "Emerging Profiles for Cultured Meat; Ethics through and as Design," *Animals* 3 (2013): 647–62, doi:10.3390/ani3030647.

144 **That's what came**: Cor van der Weele and Clemens Driessen, "How Normal Meat Becomes Stranger as Cultured Meat Becomes More Normal; Ambivalence and Ambiguity below the Surface of Behavior," *Frontiers in Sustainable Food Systems* 3 (2019): 69. doi:10.3389/fsufs.2019.00069.

REFERENCES

Abbey, Edward. *Desert Solitaire: A Season in the Wilderness* 1968. Reprint. New York: Ballentine, 1985.

Alaimo, Stacy and Hekman, Susan, eds. *Material Feminisms*. Bloomington: Indiana University Press, 2008.

Altringham, John D. *Bats: From Evolution to Conservation*. Oxford: Oxford University Press, 2011.

Auerbach, Eric. *Mimesis: The Representation of Reality in Western Thought*. 1946. Reprint. Princeton, NJ: Princeton University Press, 2016.

Barad, Karen. *Meeting the Universe Halfway: Quantum Physics and the Entanglement of Matter and Meaning*. Durham, NC: Duke University Press, 2007.

Bar-Cohen, Yoseph, ed. *Biomimetics: Nature-Based Innovation*. Boca Raton, FL: CRC Press, 2011.

———, ed. *Biomimetics: Biologically Inspired Technologies*. Boca Raton, FL: CRC Press, 2006.

Bennett, Jane. *Vibrant Matter: A Political Ecology of Things*. Durham, NC: Duke University Press, 2010.

Benyus, Janine. *Biomimicry: Innovation Inspired by Nature*. New York: William Morrow, 1997.

Bogost, Ian. *Alien Phenomenology: Or What It's Like to Be a Thing*. Minneapolis: University of Minnesota Press, 2013.

Brooks, Rodney. "Elephants Don't Play Chess." *Robotics and Autonomous Systems* 6 (1990): 3–15.

———. *Flesh and Machines: How Robots Will Change Us*. New York: Pantheon, 2002.

Bryant, Levi. *The Democracy of Objects*. Ann Arbor, MI: Open Humanities, 2011.

Brynjolfsson, Erik and Andrew McAfee. *The Second Machine Age: Work, Progress and Prosperity in a Time of Brilliant Technologies*. New York: Norton, 2014.

Coetzee, John Maxwell *Elizabeth Costello*. New York: Viking, 2003.

Coole, Diana and Samantha Frost, eds. *New Materialisms: Ontology, Agency and Politics*. Durham, NC: Duke University Press, 2010.

Cronon, William. "The Trouble with Wilderness." In *Uncommon Ground: Rethinking the Human Place in Nature*, ed. William Cronon. New York: Norton (1995), 69–90.

Damasio, Antonio. *The Strange Order of Things: Life, Feeling and the Making of Cultures.* New York: Pantheon, 2018.

———. *Self Comes to Mind: Constructing the Conscious Brain.* New York: Vintage, 2012.

———. *Looking for Spinoza: Joy, Sorrow and the Feeling Brain.* New York: Harcourt, 2003.

———. *Descartes' Error: Emotion, Reason and the Human Brain.* New York: Putnam, 1994.

Davies, Paul and Philip Clayton, eds. *The Re-Emergence of Emergence: The Emergentist Hypothesis from Science to Religion.* Oxford: Oxford University Press, 2006.

De Little, Alex. *Spatial Listening: Exercises in Aural Architecture.* London: Wild Pansy Press, 2019.

Derrida, Jacques. "The Animal That Therefore I Am (More to Follow)." *Critical Inquiry* 28, no. 2 (2002): 369–418.

Dorigo, Marco and Thomas Stützle. *Ant Colony Optimization.* Cambridge: MIT Press, 2004.

Foster, Charles. *Being a Beast: Adventures across the Species Divide.* New York: Metropolitan Books, 2016.

Gordon, Deborah M. *Ant Encounters: Interaction Networks and Colony Behavior.* Princeton, NJ: Princeton University Press, 2010.

Halliwell, Stephen. *The Aesthetics of Mimesis: Ancient Texts and Modern Problems.* Princeton, NJ: Princeton University Press, 2002.

Haraway, Donna. *Staying with the Trouble: Making Kin in the Chthulucene.* Durham, NC: Duke University Press, 2016.

———. *When Species Meet.* Minneapolis: University of Minnesota Press, 2008.

———. *Simians, Cyborgs, and Women: The Reinvention of Nature.* London: Routledge, 1990.

Harman, Graham. *Object-Oriented Ontology: A New Theory of Everything.* London: Penguin, 2018.

———. *Guerilla Metaphysics: Phenomenology and the Carpentry of Things.* Chicago, IL: Open Court, 2005.

———. *Tool Being: Heidegger and the Metaphysics of Objects.* Chicago, IL: Open Court, 2002.

Higgs, Eric. *Nature by Design: People, Natural Process, and Ecological Restoration.* Cambridge: MIT Press, 2003.

Holland, Owen and David McFarland, eds. *Artificial Ethology.* Oxford: Oxford University Press, 2001.

Hölldobler, Bert and E.O. Wilson. *The Superorganism: The Beauty, Elegance, and Strangeness of Insect Societies.* New York: Norton, 2009.

———. *The Ants.* Cambridge: Harvard University Press, 1990.

Kelleher, John D. *Deep Learning.* Cambridge: MIT Press, 2019.

Latour, Bruno. *Facing Gaia: Eight Lectures on the New Climate Regime.* Medford, OR: Polity Press, 2017.

———. *Politics of Nature: How to Bring the Sciences into Democracy.* Cambridge, MA: Harvard University Press, 2004.

Leopold. Aldo. *A Sand County Almanac.* 1949. Reprint. Oxford: Oxford University Press, 1992.

Margulis, Lynn. *Symbiotic Planet: A New Look at Evolution.* New York: Basic Books, 1999.

Markoff, John. *Machines of Loving Grace: The Quest for Common Ground between Humans and Robots.* New York: Ecco, 2015.

Maturana, Humberto and Francisco Varela. *Autopoiesis and Cognition: The Realization of the Living.* Boston, MA: Reidel, 1980.

Moravec, Hans. *Mind Children*. Cambridge, MA: Harvard University Press, 1988.

Morton, Timothy. *Hyperobjects: Philosophy and Ecology after the End of the World*. Minneapolis: University of Minnesota Press, 2013.

———. *Ecology after Nature: Rethinking Environmental Aesthetics*. Cambridge, MA: Harvard University Press, 2007.

McKibben, Bill. *The End of Nature*. New York: Random House, 1989.

Nagel, Thomas. "What Is it Like to Be a Bat?" *The Philosophical Review* 83 (1974): 435–50.

Oates, Joyce Carol. "Against Nature." *Antaeus* 57 (1986): 238–43.

Oliveros, Pauline. *Deep Listening: A Composer's Sound Practice*. Bloomington, IN: iUniverse, 2005.

Pollan, Michael. *The Omnivore's Dilemma: A Natural History of Four Meals*. New York: Penguin, 2006.

Quammen, David. *The Flight of the Iguana*. New York: Touchstone, 1988.

Rorty, Richard. *Philosophy and the Mirror of Nature*. 1979. Reprint. Princeton, NJ: Princeton University Press, 2017.

Seeley, Thomas D. *Honeybee Democracy*. Princeton, NJ: Princeton University Press, 2010.

———. *The Wisdom of the Hive: The Social Physiology of Honey Bee Colonies*. Cambridge, MA: Harvard University Press, 1995.

Sharbafi, Maziar A. and André Seyfarth, eds. *Bioinspired Legged Locomotion: Models, Concepts and Applications*. London: Butterworth-Heinemann, 2017.

Sheldrake, Merlin. *Entangled Life: How Fungi Make Our World, Change Our Minds and Make Our Futures*. New York: Random House, 2020.

Shepard, Paul. *Coming Home to the Pleistocene*. Washington, DC: Island Press, 1998.

Snyder, Gary. *The Practice of the Wild*. San Francisco, CA: North Point Press, 1990.

Taussig, Michael. *Mimesis and Alterity: A Particular History of the Senses*. London: Routledge, 1993.

Thwaites, Thomas. *Goatman: How I Took a Holiday from Being Human*. Princeton, NJ: Princeton Architectural Press, 2016.

Tsing, Anna Lowenhaupt. *The Mushroom at the End of the World: On the Possibility of Life in Capitalist Ruins*. Princeton, NJ: Princeton University Press, 2015.

Turkle, Sherry. *Alone Together: Why We Expect More from Technology and Less from Each Other*. New York: Basic Books, 2011.

Voegelin, Salomé. *The Political Possibility of Sound: Fragments of Listening*. London: Bloomsbury Academic, 2018.

———. *Listening to Noise and Silence: Towards a Philosophy of Sound Art*. London: Continuum, 2010.

Vogel, Steven. *Cats' Paws and Catapults: Mechanical Worlds of Nature and People*. New York: Norton, 1998.

Wilson, Edward O. *Biophilia*. Cambridge, MA: Harvard University Press, 1984.

Wohlleben, Peter. *The Hidden Life of Trees: What They Feel, How They Communicate*. Berkeley, CA: Greystone Books, 2015.

INDEX

For Product Safety Concerns and Information please contact our EU representative GPSR@taylorandfrancis.com Taylor & Francis Verlag GmbH, Kaufingerstraße 24, 80331 München, Germany

Batch number: 10725798

Printed by Printforce, the Netherlands